Memoirs of the American Mathematical Society
Number 361

Frank Rimlinger

Pregroups and Bass-Serre theory

Published by the
AMERICAN MATHEMATICAL SOCIETY
Providence, Rhode Island, USA

January 1987 · Volume 65 · Number 361 (fourth of 5 numbers)

MEMOIRS of the American Mathematical Society

SUBMISSION. This journal is designed particularly for long research papers (and groups of cognate papers) in pure and applied mathematics. The papers, in general, are longer than those in the TRANSACTIONS of the American Mathematical Society, with which it shares an editorial committee. Mathematical papers intended for publication in the Memoirs should be addressed to one of the editors:

PREPARATION OF COPY. Memoirs are printed by photo-offset from camera-ready copy prepared by the authors. Prospective authors are encouraged to request a booklet giving detailed instructions regarding reproduction copy. Write to Editorial Office, American Mathematical Society, Box 6248, Providence, RI 02940. For general instructions, see last page of Memoir.

SUBSCRIPTION INFORMATION. The 1987 subscription begins with Number 358 and consists of six mailings, each containing one or more numbers. Subscription prices for 1987 are $227 list, $182 institutional member. A late charge of 10% of the subscription price will be imposed on orders received from nonmembers after January 1 of the subscription year. Subscribers outside the United States and India must pay a postage surcharge of $25; subscribers in India must pay a postage surcharge of $43. Each number may be ordered separately; *please specify number* when ordering an individual number. For prices and titles of recently released numbers, see the New Publications sections of the NOTICES of the American Mathematical Society.

BACK NUMBER INFORMATION. For back issues see the AMS Catalogue of Publications.

Subscriptions and orders for publications of the American Mathematical Society should be addressed to American Mathematical Society, Box 1571, Annex Station, Providence, RI 02901-9930. *All orders must be accompanied by payment.* Other correspondence should be addressed to Box 6248, Providence, RI 02940.

MEMOIRS of the American Mathematical Society (ISSN 0065-9266) is published bimonthly (each volume consisting usually of more than one number) by the American Mathematical Society at 201 Charles Street, Providence, Rhode Island 02904. Second Class postage paid at Providence, Rhode Island 02940. Postmaster: Send address changes to Memoirs of the American Mathematical Society, American Mathematical Society, Box 6248, Providence, RI 02940.

Printed in the United States of America.

The paper used in this journal is acid-free and falls within the guidelines established to ensure permanence and durability. ∞

Table of Contents

Abstract

This paper investigates the structure of pregroups, Stallings generalization of free products with amalgamation. Stallings discovered an order relation on pregroups which gives rise to the concept of *finite height*. We show that the universal group of a pregroup P of finite height may be realized as the fundamental group of a graph of groups. The vertex groups of this graph of groups correspond to the P-conjugacy classes of the maximal subgroups of P. Conversely, we construct a pregroup structure for the fundamental group of any graph of groups whose geodesics are of finite bounded length. Via a theorem of Karrass, Pietrowski, and Solitar, we deduce that a group is a finite extension of a finitely generated free group if and only if it is the universal group of a finite pregroup.

1980 *Mathematics Subject Classification*. Primary 20E07, 20E34, 20F10; Secondary 20E06.

Keywords and phrases. Pregroup, word problem, reduced word, graph of groups, HNN extension, free by finite.

Library of Congress Cataloging-in-Publication Data

Rimlinger, Frank, 1957–
Pregroups and Bass-Serre theory.

(Memoirs of the American Mathematical Society, 0065-9266;
no. 361 (Jan. 1987))
"January 1987, volume 65, number 361 (fourth of 5 numbers)."
Bibliography: p.
1. Pregroups. I. Title. II. Title: Bass-Serre theory. III. Series.
QA3.A57 no. 361 [QA171] 510 s [512′.22] 86-32112
ISBN 0-8218-2421-X

Introduction

Free groups, free products with amalgamations, and HNN extensions may be thought of as "groups defined by reduced words." We make this concept precise as follows: Let G be a group, and let $P \subset G$ be a subset of G such that P generates G and $P=P^{-1}$. Let $D \subset P \times P$ be a subset of pairs (x,y) of elements of P such that for all $(x,y) \in D$, $xy \in P$. A word $(x_1, \ldots, x_n) \in P^n$ is said to be *reduced* if for each adjacent pair (x_i,x_{i+1}) it is *not* the case that $(x_i,x_{i+1}) \in D$, (for otherwise we reduce the word in the obvious way). We say that (P,D) is a *reduced word structure* for G if for all $g \in G$ all reduced words representing g are of the same length.

In Rimlinger [to appear] we proved that if (P,D) is a reduced word structure for G, then (P,D) is a pregroup and G is isomorphic to $\mathbf{U}(P)$, the universal group of P. But what is a pregroup? These objects were created by John Stallings [1971], who traces his idea back to van der Waerden [1948] and Baer [1950]. In Stallings treatment, a pregroup is a set P, together with a subset $D \subset P \times P$, and a special element $1 \in P$, called the identity element. In addition, pregroups are endowed with a *partial multiplication* $m:D \to P$ and an involution $i:P \to P$. Moreover, certain axioms concerning the above sets and maps must also hold. These axioms may be paraphrased by saying that 1 is the identity element, $m:D \to P$ is "as associative a multiplication as possible," and $i:P \to P$ takes an element to its inverse. Additionally, if three pairs (w,x), (x,y), and (y,z) are in D, then either $(wx,y) \in D$ or $(xy,z) \in D$.

Stallings defined the *universal group* $\mathbf{U}(P)$ of a pregroup (P,D) to be the free group on the set P modulo the relations $\{ xy=z \mid x,y,z \in P \text{ and } z=m(x,y) \}$. If G is a group isomorphic to $\mathbf{U}(P)$, we say, somewhat loosely, that G has a *pregroup structure* P. Stallings original theorem about pregroups, interpreted in the language developed above, is that (P,D) is a reduced word structure for $\mathbf{U}(P)$. Thus pregroups exactly capture the notion of "groups defined by reduced words," in that for any group G, (P,D) is a reduced word structure for G if and only if (i) (P,D) is a pregroup, (ii) G and $\mathbf{U}(P)$ are isomorphic, and hence (iii) P is a pregroup structure for G.

For example, if $P = \{ x,x^{-1},1 \}$,

$$D = \{ (x,x^{-1}),(x^{-1},x),(1,x),(x,1),(1,x^{-1}),(x^{-1},1),(1,1) \}$$

and $m:D \to P$ and $i:P \to P$ are defined in the obvious way, then (P,D) is in fact a pregroup, and $\mathbf{U}(P)$ is a free group of rank 1. The reduced words with respect to D representing elements of $\mathbf{U}(P)$ are obtained from arbitrary words in the alphabet P via free reduction.

Stallings showed that free products with amalgamation and HNN extensions also have pregroup structures. It is a fact that if (P,D) is a pregroup, $Q \subset P$, $Q \times Q \subset D$, and $(x,y) \in Q \times Q \Rightarrow xy \in Q$, then Q inherits a group structure from P. In this event we say that Q is a *subgroup of* P. For example, in Stallings's original constructions of pregroup structures for $A *_C B$ and $A \circlearrowleft_C$, the maximal subgroups of the pregroup structures are A and B in the case of the free product with amalgamation and A in the HNN case.

Two natural questions arise at this point. Given a pregroup P, does $\mathbf{U}(P)$ act on a tree? Conversely, given a group which acts on a tree, does this group have a pregroup structure which reflects the structure of the action. We answer both these questions in the affirmative, provided (i) the tree in question is an ordinary simplicial tree, (ii) the quotient graph of the tree in question has geodesics of finite bounded length, and (iii) the pregroup in question satisfies pregroup theoretic criteria analogous to (i) and (ii). The exact results are stated here. The summary below refers the pregroup theoretic notions to the main body of the paper.

Theorem A: Let P be a pregroup of finite height . Let $\mathbf{G}$ be the union of the fundamental group systems for each $U^i(P)$, $i=0, \ldots, d$. Then P is a pregroup structure for the fundamental group of a graph of groups $\pi_1(\mathbf{H},Y)$. The base groups of Y are in 1-1 correspondence with the elements of $\mathbf{G}$, and the corresponding groups are isomorphic. The graph Y is such that the oriented edges of the complement of a maximal tree in Y are in 1-1 correspondence with the union of the spanning sets of the fundamental groups of $\mathbf{G}$.

Theorem B: Let $(\mathbf{H},Y)$ be a graph of groups with bases V, edges E, basepoint v_0, and maximal tree T. Suppose the graph Y is of finite diameter and $(\mathbf{H},Y)$ is proper. Let $X \subset E$ be the edges of Y not in T, and let X^+ be an orientation of X. Then $\pi_1(\mathbf{H},Y,v_0)$ has a pregroup structure Q satisfying the following conditions:

(i) for some $d \geq 0$, Q has depth d, and a good Q-sequence $((\mathbf{G}_d,\varnothing), \ldots, (\mathbf{G}_1,\varnothing),(\mathbf{G}_0, \tilde{X}))$ satisfying (ii), (iii), and (iv):

(ii) for $i=1, \ldots, d$, the groups of $\mathbf{G}_i$ are in 1-1 correspondence with the maximal bases of depth i, and corresponding fundamental groups and base groups are isomorphic,

(iii) $\tilde{X}^+$ is in 1-1 correspondence with X^+,

(iv) $\mathbf{G}_0$ is in 1-1 correspondence with

$\{ v \in B \mid \text{depth}(v)=0 \} \cup \hat{B}$, where $\hat{B} \subset \{ v \in B \mid \text{depth}(v) > 0 \}$.

If $G \in \mathbf{G}_0$ corresponds to G_v for some $v \in B$ of depth 0, then G is isomorphic to $H_v \in \mathbf{H}$. If $G \in \mathbf{G}_0$ corresponds to some $v \in \hat{B}$, then G is isomorphic to a subgroup of H_v. □

From theorems A and B and result of Karrass, Pietrowski, and Solitar [1973] generalizing Stallings' work on the ends of a group [1968], we deduce the following corollary.

Corollary: A group G is free by finite if and only if G is the universal group of a finite pregroup.

The paper is organized as follows:

Part I: We review Stallings' work on pregroups and the connection between pregroups and groups defined by reduced words. We make an initial investigation of the *subgroups* of a pregroup. We provide a geometric interpretation of the pregroup axioms, which serves as a convenient computational tool for part II.

We exploit a *tree ordering,* discovered by Stallings, on the elements of a pregroup in order to define the subcategory of pregroups of *finite height.* We define the *full subpregroup of units,* $U(P)$, of a pregroup P, and show how this leads to a descending sequence $P \supset U(P) \supset U^2(P) \cdots \supset U^d(P)$ of subpregroups of given pregroup P of finite *depth* d.

Part II: We define the notion of a *pregroup action* on a set. We show that the subpregroup of units $U(P)$ acts on the *maximal vertices* of P. We exploit this action to define a *generating set* for P, consisting of a *fundamental group system* and a *spanning set* for P. We prove that in a certain sense $U(P) \cup \{$ fundamental group system $\} \cup \{$ spanning sets $\}$ is a subset of P which minimally generates $\mathbf{U}(P)$, (theorem 4.24). This result paves the way for theorem 5.3, in which we give a presentation of P in terms of a generating set of P. We give an example which illustrates this presentation and shows how to get from a pregroup to a graph of groups. As an easy application of this presentation, we prove the the universal group of a finite pregroup with no nontrivial subgroups is free.

Part III: We give an example which indicates the major technical problem of the proof of theorem A. This example motivates an inductive argument based on the presentation of part II. To prove theorem B, we first review chapter one of Serre [1980] from a pregroup theoretic point of view. We construct a pregroup structure, due to Stallings, for $F(\mathbf{H},Y)$. This group is defined in Serre [1980] and is a large group containing $\pi_1(\mathbf{H},Y,v_0)$, the fundamental group of a graph of groups. We define, for each $\alpha \in F(\mathbf{H},Y)$, a sequence of paths $Y(\alpha)$ in Y reflecting the reduced word structure for α, which in turn is derived from the pregroup structure for $F(\mathbf{H},Y)$. Using this idea we define a subset Q of $F(\mathbf{H},Y)$ and prove that Q is in fact a subpregroup of $F(\mathbf{H},Y)$. Finally, by reference to the proof of theorem A, we show that $\mathbf{U}(Q)$ is isomorphic to $\pi_1(\mathbf{H},Y,v_0)$, establishing theorem B. We end the paper with examples showing how to go from a free product with amalgamation or an HNN extension to a pregroup. These examples resemble some of the original examples of Stallings [1971].

I would like to thank John Stallings for helping me get started in the pregoup business, and for many valuable suggestions. In particular, the suggestion that pregroups were somehow related to Bass-Serre theory set me going in the right direction.

Part I
Stallings's theorem about pregroups; introduction to the pregroups of finite height

1. The definition of a pregroup

In this section we review some basic facts about pregroups. We define reduced word structures and pregroup structures. By theorem 3, theorem 4 and corollary 7 below, these concepts are equivalent. References are given for proofs of theorems 3 and 4, but otherwise the exposition is self contained. Anticipating results to come, we end this section with a discussion of subgroups of a pregroup.

We start by considering *prees*, or sets with partial multiplication table. Let P be a set, let $D \subset P \times P$, and let $m: D \to P$ be a set map. Typically we denote $m(x,y)$ by xy or $x \bullet y$. The composite concept (P,D,m) is called a *pree*. The *universal group* $\mathbf{U}(P)$ of a pree P is determined by the presentation $(P; \{ m(x,y)y^{-1}x^{-1} \mid (x,y) \in D \})$. The prees form a category $\mathbf{C}$. Given prees (P,D) and (Q,E), we define $[P,Q]_{\mathbf{C}}$ to be the set maps $\phi: P \to Q$ such that $(x,y) \in D \Rightarrow (\phi(x),\phi(y)) \in E$ and $\phi(xy) = \phi(x)\phi(y)$. Clearly there is a forgetful functor $\mathbf{F}: \mathbf{G} \to \mathbf{C}$ from groups to prees, and if G is a group, then the diagram

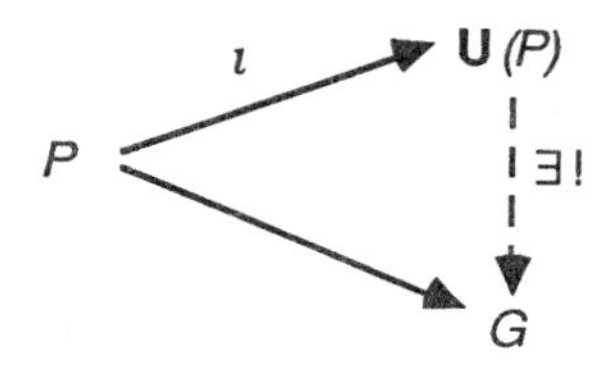

shows that $[\mathbf{U}(P),G]_{\mathbf{G}}$ and $[P,\mathbf{F}(G)]_{\mathbf{C}}$ are in 1-1 natural correspondence. The map $\iota: P \to \mathbf{U}(P)$ is the restriction to P of the natural projection $F(P) \to \mathbf{U}(P)$ from the free group on the set P onto $\mathbf{U}(P)$. Let (P,D) be a pree. Here is some useful terminology.

P-word X of *length* n: An element $X = (x_1, \ldots, x_n) \in P^n$

X is P*-reduced*: This means $\forall\ 1 \le i \le n-1,\ (x_i, x_{i+1}) \notin D$.

X *represents* x: This means $x = \iota(x_1)\iota(x_2) \cdots \iota(x_n) \in \mathbf{U}(P)$.

Recieved by the editor July 14, 1985.

*The author was a Sloan Foundation Doctoral Dissertation Fellow at UC Berkeley during the period this paper was written.

$(x_1, \ldots, x_n)_D$: This means $\forall\ 1 \leq i \leq n-1$, $(x_i, x_{i+1}) \in D$.

(w,x,y) *associates*: This means (i) (w,x,y) is a P-word, and (ii) $(w,x,y)_D$, $(w,xy)_D$, and $(wx,y)_D$.

Suppose $X=(x_1, \ldots, x_n)$ is a P-word and $(x_i,x_{i+1})_D$ for some $1 \leq i < n$. Then $Y=(x_1, \ldots, x_i x_{i+1}, \ldots, x_n)$ is an *elementary* reduction of X. Clearly X and Y represent the same element of $\mathbf{U}(P)$. If $g \in \mathbf{U}(P)$ is represented by some P-word X, then by making elementary reductions we may proceed from X to a P-reduced word Y representing g. It is interesting to note that some elements of $\mathbf{U}(P)$ may not have any representation at all because P does not contain the "inverse elements." A more serious objection to the category of prees is that the map $\iota: P \to \mathbf{U}(P)$ is not injective in general. The fatal blow for $\mathbf{C}$ is that there is no general algorithm for deciding when two arbitrary P-words represent the same element of $\mathbf{U}(P)$. We now introduce Stallings's category of pregroups, where all these objections disappear.

1. Definition: Let (P,D) be a pree. Suppose P contains a distinguished element $1 \in P$, and is endowed with an involution $i: P \to P$, denoted $i(x)=x^{-1}$. Then (P,D) is a *pregroup* if it satisfies properties (P1) to (P4) for all $w,x,y,z \in P$.

(P1): $(1,x)_D, (x,1)_D$ and $1x=x=x1$.

(P2): $(x^{-1},x)_D$, $(x,x^{-1})_D$, and $xx^{-1}=x^{-1}x=1$.

(P3): Suppose $(w,x,y)_D$. Then $\{\ (wx,y)_D$ or $(w,xy)_D\ \} \Rightarrow \{\ (w,x,y)$ associates and $(wx)y=w(xy)\ \}$.

(P4): If $(w,x,y,z)_D$ then $\{\ (w,xy)_D$ or $(xy,z)_D\ \}$. □

Axioms (P1) and (P2) are transparent, and (P3) captures the idea of "associativity whenever possible." In Rimlinger [to appear] there is an alternative characterization of pregroups in terms of "groups derived from reduced words," as follows.

2a. Definition: Suppose P is a pregroup, G is a group, $P \subset G$, and the inclusion map $P \to G$ is a pregroup morphism. Let $\phi: \mathbf{U}(P) \to G$ be the induced group homorphism. Then P is a *pregroup structure* for G if $\phi: \mathbf{U}(P) \to G$ is an isomorphism.

2b. Definition: Let G be a nontrivial group. Suppose $P \subset G$, $P=P^{-1}$, and P generates G. Suppose $D \subset P \times P$; and suppose that $(x,y) \in D$ implies $xy \in P$. Clearly (P,D) is a pree. By abuse of notation, the P-words represent elements of G. Suppose all the P-reduced words representing the same element of G are of the same P-length. Then (P,D) is called a *reduced word structure* for G.

3. Theorem: (Rimlinger [to appear]) If P is a reduced word structure for G, then P is a pregroup structure for G. □

Stallings' original theorem about pregroups implies the converse of theorem 3. Stallings' result is harder to prove and deeper, in that it provides precise information about the word problem in $\mathbf{U}(P)$. In fact if P is a *finite* pregroup, then Stallings' theorem implies $\mathbf{U}(P)$ has solvable word problem.

Suppose P is a pregroup. Let $X=(x_1, \ldots, x_n)$ and $A=(1=a_0,a_1, \ldots, a_{n-1},a_n=1)$ be P-words. Suppose for $i=1, \ldots, n$ that (a_{i-1}^{-1},x_i,a_i) associates. We then define a P-word $X \bullet A = (x_1a_1,a_1^{-1}x_2a_2, \ldots, a_{n-1}^{-1}x_n)$ called X *interleaved by* A. Let R be the set of reduced words of P. Given P-reduced words $X,Y \in R$, we say $X \sim Y$ if $X \bullet A = Y$ for some P-word A. The relation $\sim$ is an equivalence relation on R, and each $\sim$-class of R represents a unique element of $\mathbf{U}(P)$. Conversely, each element of $\mathbf{U}(P)$ is represented by at least one $\sim$-class. But much more is true.

4. Theorem: (Stallings [1971, Theorem 3.A.4.5]) Let (P,D) be a pregroup. If X,Y are P-reduced words, then X and Y represent the same element of $\mathbf{U}(P)$ if and only if $X \sim Y$. $\square$

In the light of theorem 4, we can make the following definition of the length of an element of $\mathbf{U}(P)$.

5. Definition: Let P be a pregroup and let $x \in \mathbf{U}(P)$. Then the P-length of x, denoted $l_P(x)$, is the length of a reduced word representing x. In the case of $1 \in P$, we insist that $l_P(1)=0$, even though the set of reduced words of length 0 is empty.

The following corollaries of this theorem are worth examining. Let (P,D) be a pregroup. Let (P,D) be a pregroup.

6. Corollary: $\iota : P \rightarrow \mathbf{U}(P)$ is injective.

Henceforth we identify $x \in P$ with $\iota(x) \in \mathbf{U}(P)$. In particular, if $(x,y) \in P \times P$, then xy is in $\mathbf{U}(P)$ but perhaps not in P.

7. Corollary: P is a reduced word structure for $\mathbf{U}(P)$.

Thus theorem 3 may be viewed as an alternative definition for pregroups.

8. Corollary: $(x,y) \in D \iff \{ (x,y) \in P \times P \text{ and } xy \in P \}$

$\iff \{ (x,y) \in P \times P \text{ and } l_P(xy) \leq 1 \}$.

9. Corollary: (P,D) is a group if $D = P \times P$, because in this case theorem 4 implies that $\mathbf{U}(P) \approx P$. $\square$

We end this section with a discussion of the subgroups of a pregroup.

9a. Definition: Let (P,D) be a pregroup. A *subpregroup* of P is a pregroup (Q,E) such that $Q \subset P$, $E \subset D$, and identity, inversion, and multiplication are inherited from P. A *subgroup* of P is a subgroup Q of $\mathbf{U}(P)$ whose trivial pregroup structure $(Q, Q \times Q)$ is a subpregroup of P. $\square$

As in corollary 9, we see that a subpregroup (Q,E) of (P,D) is a subgroup of (P,D) $\Leftrightarrow$ $E = Q \times Q \subset D$. A slightly less obvious fact is

10. Corollary: Let $x \in P$. The following are equivalent: (i) x is in a subgroup of P, (ii) $x^2 \in P$, (iii) $(x,x) \in D$, (iv) Let $<x>$ be the cyclic subgroup generated by x in U(P). Then $<x>$ is a subgroup of P.

Proof: Use corollary 8 to show (i) $\Rightarrow$ (ii) $\Rightarrow$ (iii). Since (iv) implies (i), it suffices to prove (iii) $\Rightarrow$ (iv). Suppose $(x,x)_D$. Then $x^2 \in P$. Inductively, suppose $x^i \in P$ for all i, $1 \leq i \leq n$. Then $(x^{n-1},x)_D$ and (x,x^{n-1}) by corollary 8, so $(x,x^{n-1},x,x)_D$, whence axiom (P4) implies $x^{n+1} \in P$. From (P2) we see that $<x> \subset P$. Thus $<x>$ is a subgroup of P by corollaries 8 and 9. □

The elements of a pregroup fall naturally into two sets: those contained in subgroups of P and those not contained in subgroups of P.

11. Definition: Let P be a pregroup. An element $x \in P$ contained in a subgroup of P is called a *cyclic* element of P. If $x \in P$ is not cyclic, then it is *simple*. A pregroup with no non-trivial cyclic element is called a *simple pregroup*.

2. The pregroups of finite height

In this section we define the tree ordering for pregroups discovered by Stallings. In certain cases, this tree ordering determines an actual tree. In order to prove that pregroups have a tree ordering, and in order to prove many other results lying further down the road, we introduce a geometric method of reasoning about pregroups. This method is based on the Cayley graph of $\mathbf{U}(P)$ with colors in P. At the end of this section we define the pregroups of finite height. Eventually we shall determine the structure of the universal groups of these pregroups, by proving theorems A and B stated in the introduction.

We begin our discussion by considering tree orderings on arbitrary sets.

1. Definition: (Stallings [1971, example 3.A.5.2]) Suppose $(P, \lesssim)$ is a set with a relation that is transitive and reflexive. Then $(P, \lesssim)$ is a *tree ordering* if it has a smallest element 1 and $\forall\ x,y,z \in P$, $\{ x \lesssim z$ and $y \lesssim z \} \Rightarrow \{ x \lesssim y$ or $y \lesssim x \}$. □

As usual, if $\{ x \lesssim y$ or $y \lesssim x \}$, we say that x and y are *comparable.* If both $x \lesssim y$ and $y \lesssim x$ we say $x \approx y$. If $x \lesssim y$ but not $y \lesssim x$ we say $x < y$. An element $x \in P$ is of *finite height* if $\exists\ n \geq 0$ and an ordered subset $(x_0, \ldots, x_n)$ of P such that $1 = x_0 < x_1 < \cdots < x_n = x$ and for each $0 \leq i \leq n-1$, if $z \in P$ and $x_i \lesssim z \lesssim x_{i+1}$ then $x_i \approx z$ or $z \approx x_{i+1}$. Suppose every element of the given order tree P is of finite height. We can then turn P into an actual tree T, as defined in Serre [1980]. The vertices of T are the $\approx$-classes of P. Thus each element $x \in P$ represents a vertex V_x of T. The positively oriented edges of P are pairs (V_x, V_y) such that $x<y$ and $\forall\ z \in P$, $x \lesssim z \lesssim y \Rightarrow \{ x \approx z$ or $y \approx z \}$. Given an edge $(V_x, V_y) \in T$, the opposite edge is (V_y, V_x).

2. Lemma: T is a tree.

Proof: T is connected since by definition there is a path from V_1 to any vertex of T. Let $E = (e_1, \ldots, e_n)$ be the oriented edges of a path p in T such that for all i such that $1 \leq i \leq n-1$ it follows that e_i and e_{i+1} are not opposite edges. Such a path is said to be *without backtracking,* Our task is to show that every path without backtracking has distinct initial and terminal vertices, or in other words, to show such paths are not loops. Fix the notation $e_1 = (V_x, V_y)$ and $e_2 = (V_y, V_z)$.

Claim: $x<y \Rightarrow y<z$. Proof: If $x<y$ and $z<y$, then $x \lesssim z$ or $z \lesssim x$. If $x \lesssim z$, then $\{ x \lesssim z$ and $z<y \} \Rightarrow x \approx z$. If $z \lesssim x$, then $\{ z \lesssim x$ and $x<y \} \Rightarrow z \approx x$. In any event, $V_x = V_z$, a contradiction, since e_1 and e_2 are not opposite edges.

Using the claim inductively, we see that if e_1 is positively oriented, so are $e_2, \ldots, e_n$. In particular, p is not a loop. If all the edges in E are negatively oriented, P is clearly not a loop. If, on the contrary, E contains both positive and negative edges, then $\exists\, k$, $1 \le k \le n-1$, such that $e_1, \ldots, e_k$ are negatively oriented and $e_{k+1}, \ldots, e_n$ are positively oriented. Suppose p is a loop. Choose $u,v,w,x,y,z \in P$ as indicated below. Then $w<u$ and $y<u$, so that $w \lesssim y$ or $y \lesssim w$. Since e_k and e_{k+1} are edges, $w \approx y$, contradicting the fact that e_k and e_{k+1} are not opposite edges. Thus p is not a loop, and therefore T is a tree. □

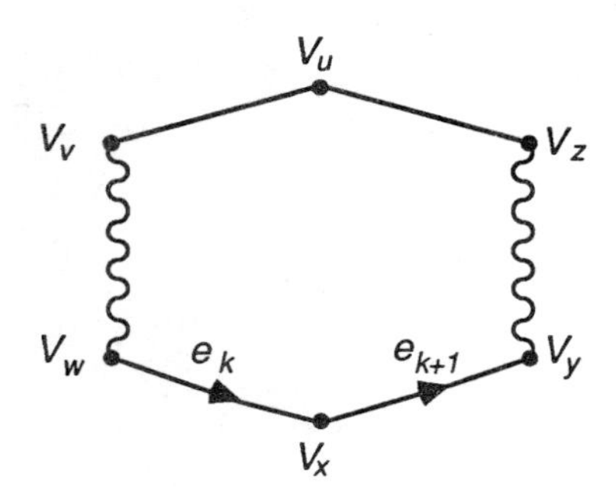

Let $x \in P$ be of finite height, and let $(x_0, \ldots, x_n)$ be as in the definition of finite height, so that $1 \approx x_0 < \cdots < x_n \approx x$ Evidently each $(V_{x_i}, V_{x_{i+1}})$ is an edge of T. It follows that n is the length of the unique geodesic from V_1 to V_x in T.

3. Definition: Let $(P, \lesssim)$ be an order tree, and suppose $x \in P$ is of finite height. Then the *height* of x, or of V_x, is the length of the geodesic from V_1 to V_x in T. In particular, V_1 has height 0. □

Pregroups and order trees are related via the following result of Stallings.

4. Definition: Let (P,D) be a pregroup. Define $\lesssim$ on P by $x \lesssim y \Leftrightarrow \{\forall\, z \in P, (z,y)_D \Rightarrow (z,x)_D\}$. Then $\lesssim$ is the *tree ordering* of P. □

We prove that $\lesssim$ is a tree ordering in theorem 8 below. First we make some general observations concerning this relation.

5. Lemma: Let $x,y \in P$. Then $x \lesssim y \Rightarrow y^{-1}x \in P \Rightarrow \{x^{-1}y \in P$ and $y^{-1}x \in P\}$.
Proof: Suppose $x \lesssim y$. The by axiom (P2), $(y^{-1},y)_D$, so that $y^{-1}x \in P$. Hence $x^{-1}y \in P$. □

6. Lemma: $x^{-1}y \in P \Leftrightarrow x$ and y are comparable.

Proof: We must verify the $(\Rightarrow)$ case. Suppose that $x^{-1}y \in P$ and that it is *not* true that $y \lesssim x$. Then there exists w such that $wy \notin P$ but $wx \in P$. Suppose $z \in P$ is such that $zy \in P$. Then $(zy, y^{-1}, x, x^{-1}w^{-1})_D$, so that by axiom P4, $zx \in P$ or $y^{-1}w^{-1} \in P$. But $y^{-1}w^{-1} \in P \Rightarrow wy \in P$, contradicting the definition of w. Thus $zx \in P$. Hence for all $z \in P$, $zy \in P$ implies $zx \in P$, or in other words, $x \lesssim y$. We have demonstrated that not $y \lesssim x$ implies $x \lesssim y$, which means that x and y are comparable. □

7. Corollary: $x \in P$ is cyclic $\Leftrightarrow$ x and x^{-1} are comparable $\Leftrightarrow$ $\exists\, z \in P$ such that $\{ x \lesssim z \text{ and } x^{-1} \lesssim z \}$.

8. Theorem: (Stallings [1971, example 3.A.5.2]) Let (P,D) be a pregroup. Then the relation $\lesssim$ defined above is a tree ordering of P.

Proof: Clearly $\lesssim$ is reflexive and transitive and $1 \in P$ is a smallest element. Let $a,b,c \in P$ and suppose $a \lesssim c$ and $b \lesssim c$. Suppose $\exists\, x \in P$ such that $xa \in P$ but $xb \notin P$. In other words, b is not $\lesssim a$. Suppose that a is not $\lesssim b$, so $\exists\, y \in P$ such that $yb \in P$ and $ya \notin P$. Since a and b are $\lesssim c$, it follows from lemma 5 that $c^{-1}a \in P$ and $c^{-1}b \in P$, and that xc and yc are *not* in P. By *fig.* 9, which will be elaborated on below, we see that xb or ya is in P, a contradiction. Thus a and b are $\lesssim$-comparable, as desired.

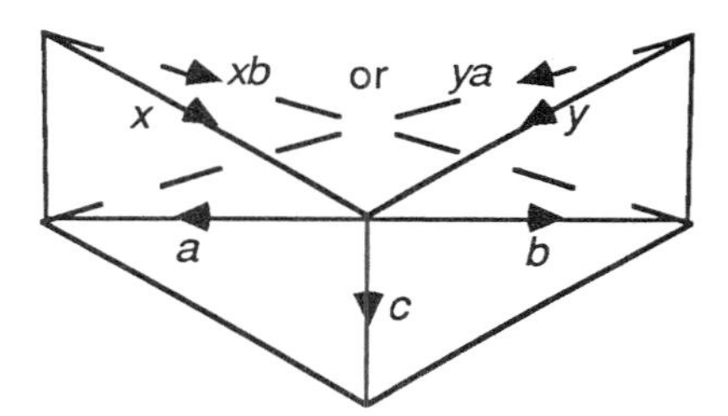

9. Figure: $a,b \leq c \Rightarrow a$ and b are comparable.

In order to see what *fig.* 9 is and why it constitutes a proof of theorem 8, we offer the following geometric interpretation of the pregroup axioms. Fix a pregroup P. Let $F(P)$ be the free group on the *set* P, and let K be the kernel of the natural projection $F(P) \to \mathbf{U}(P)$. Let R be the one vertex graph with oriented edges consisting of the set P, and let $\rho: C \to R$ be the covering projection of R corresponding to the subgroup K, so that $\rho_* \pi_1(R) = K$ in a natural way. We think of C as a graph in category of graphs defined by Serre [1980, chap. 1.2]. Specifically, C corresponds to the graph $\Gamma(G,S)$ defined near the end of section 1.2.1 of this book, with $G = \mathbf{U}(P)$ and $S = P$. The graph $\Gamma(G,S)$ is the Cayley graph G with colors in S.

Notice the vertices of C are in 1-1 correspondence with the left cosets of K in $F(P)$ and hence with the elements of $\mathbf{U}(P)$. We identify vertices of C with elements of $\mathbf{U}(P)$ via this correspondence. Given an edge $e = \overset{x \qquad y}{\bullet\!\!\longrightarrow\!\!\bullet}$ in C with initial vertex $x \in \mathbf{U}(P)$ and terminal vertex $y \in \mathbf{U}(P)$, the definition of C implies that $x^{-1}y \in P$, and we say $x^{-1}y$ is the *label* of e. Clearly given an element $x \in P$, the edges of C with label x constitute the orbit determined by $\underset{1 \qquad x}{\bullet\!\!\overset{x}{\longrightarrow}\!\!\bullet}$ as $\mathbf{U}(P)$ acts on the edges of C. This edge orbit is denoted by $[\bullet\!\!\overset{x}{\longrightarrow}\!\!\bullet]$. The point of saying all this is that when we

draw a picture such as

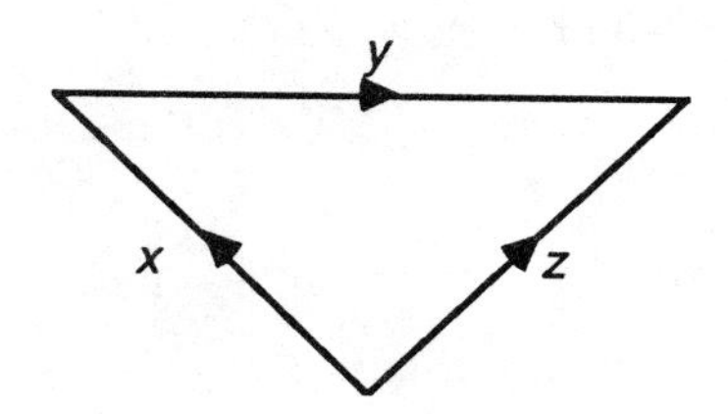

we are asserting the following logical statement:

(G1) $x, y, z \in P$, and

(G2) there is a graph morphism

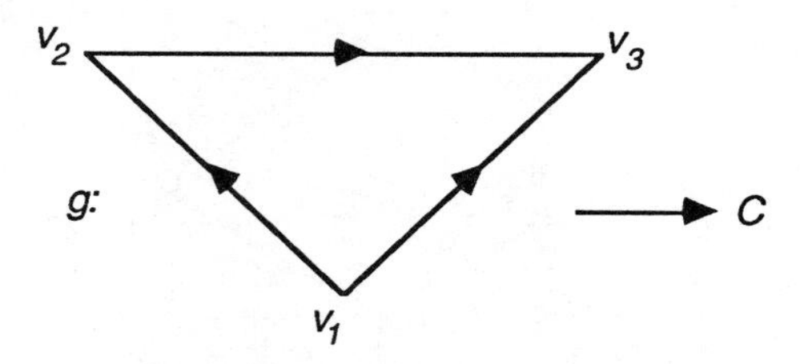

such that $g(v_1, v_2) \in [\bullet \xrightarrow{x} \bullet]$, $g(v_2, v_3) \in [\bullet \xrightarrow{y} \bullet]$, and $g(v_1, v_3) \in [\bullet \xrightarrow{z} \bullet]$. □

Remarks:

(i) In particular, (G2) does *not* imply that $g(v_1)$, $g(v_2)$, and $g(v_3)$ are distinct vertices of C.

(ii) However, we have $\Leftrightarrow xy = z \Leftrightarrow x^{-1}z = y \Leftrightarrow$ etc.

(iii) Sometimes we write if we are not interested in naming one of the edges. Notice $\Leftrightarrow xy \in P \Leftrightarrow y^{-1}x^{-1} \in P$.

(iv) This definition generalizes to arbitrary graphs Γ and maps $g:\Gamma \rightarrow C$ in the obvious way.

Let us consider observation (i) in more detail. Sometimes it is advantageous to know when vertices are distinct. Therefore a picture such as

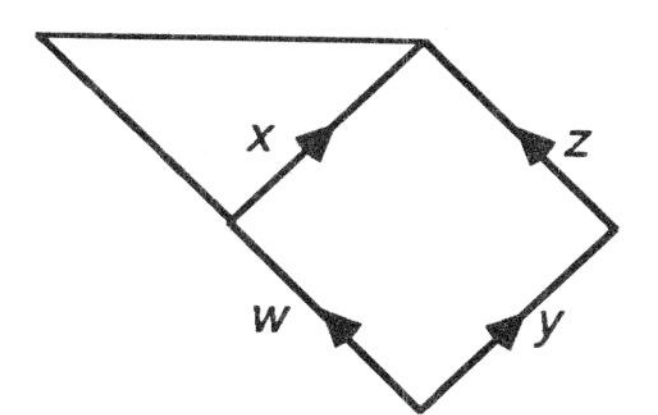

means that $\exists\ g:\Gamma\to C$ satisfying (G1) and (G2) as indicated:

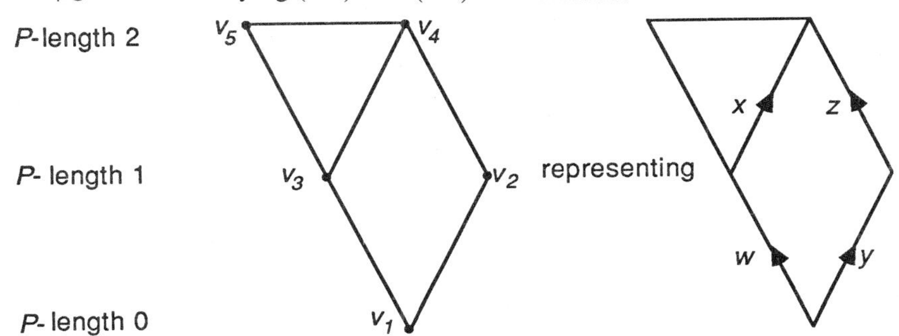

In addition, $g:\Gamma\to C$ satisfies

(G3) $g(v_1)=1$, so that $l_P(g(v_1))=0$ by definition. Moreover, $l_P(g(v_2))=l_P(g(v_3))=1$, and $l_P(g(v_4))=l_P(g(v_5))=2$.

In general, if vertices $v_i, v_j \in \Gamma$ are at the same *horizontal* level then their P-lengths $g(v_i)$ and $g(v_j)$ coincide. Moreover, the P-length of a particular vertex $g(v_i) \in C$ is well determined since the "bottom" vertex $g(v_1)$ is 1. Henceforth all the pictures we draw will have a unique bottom vertex and easily distinguished horizontal levels.

For example, from

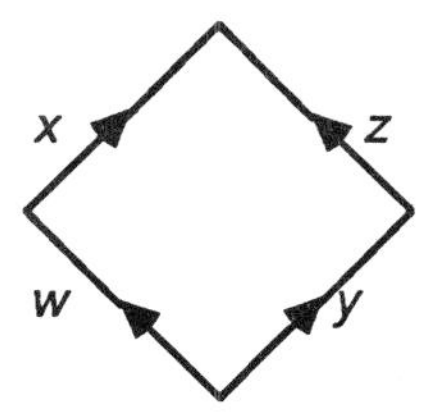

we deduce that $w \neq wx$, since $l_P(w)=1\neq 2=l_P(wx)$. However, it may still be true that $w=y$.

Finally, by virtue of ignorance or indifference, we may not want to commit ourselves to the exact P-length of some vertex. Thus

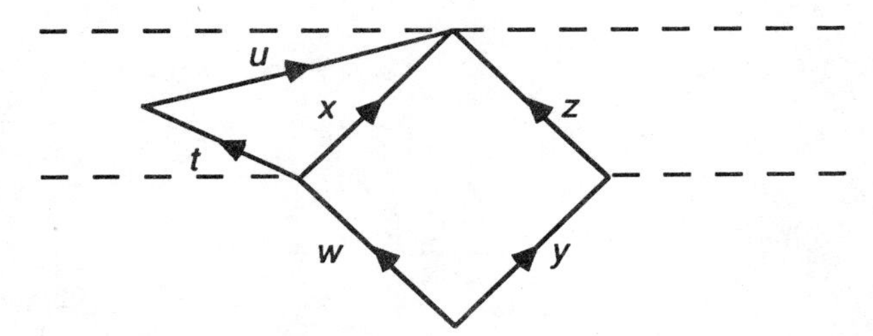

would indicate that $l_P(wt) \in \{ 1,2 \}$ but we are not sure whether the edge labeled t goes up or sideways. In any event $x = tu$ and $l_P(w(tu)) = l_P(wx) = 2$.

10. Lemma:

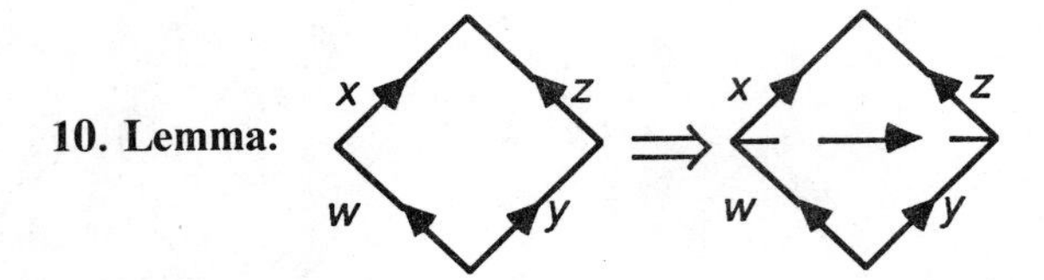

Algebraically: if $\{ w,x,y,z \in P$ such that $wx = yz$ and $l_P(wx) = 2 \}$ then $w^{-1}y \in P$. We shall go back and forth freely between the pictures and the algebra.

Proof: From Stallings's theorem 1.4, we see that the P-reduced words (w,x) and (y,z) must differ by an interleaved product. □

11: Lemma:

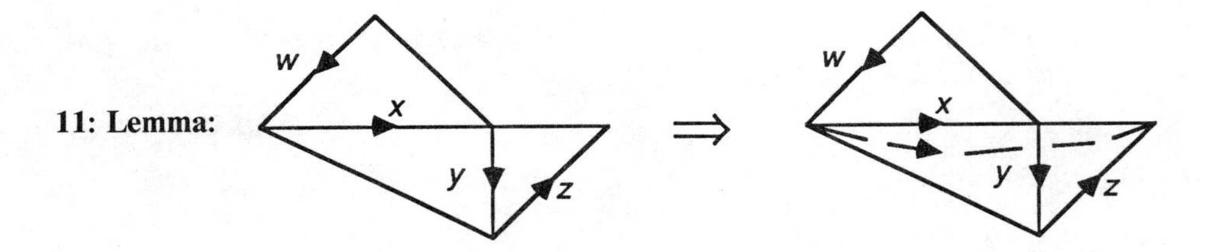

Proof: Recall the useful terminology of section 1. Clearly $(w,x,y,z)_D$, so that axiom (P4) implies $wxy \in P$ or $xyz \in P$. But $l_P(wxy) = 2$ so we must have $xyz \in P$. □

12. Lemma:

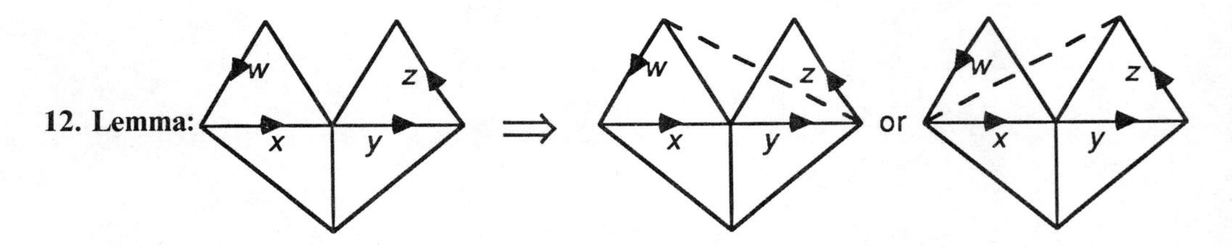

Proof:

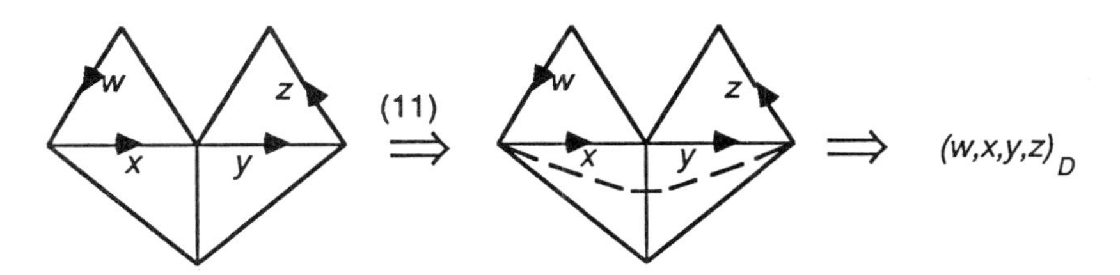

Now apply (P4). □

It is now plain that figure 9, the point of departure for this discussion, is actually a proof of theorem 8. Here are two useful lemmas which yield a geometric interpretation of $x < y$.

13. Lemma:

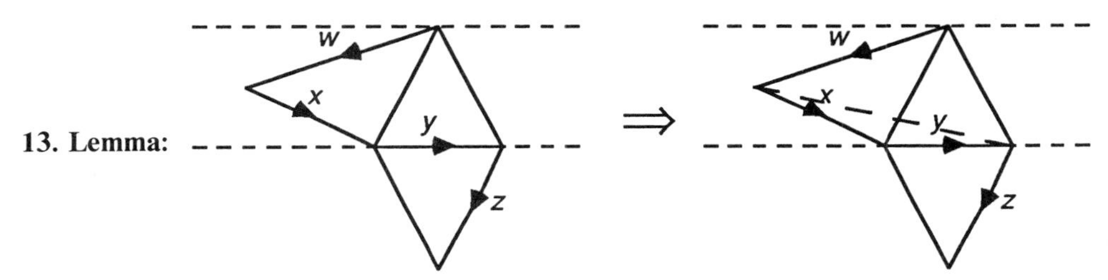

In other words, the conclusion is $xy \in P$.

Proof: Case 1: $xyz \in P$. Then

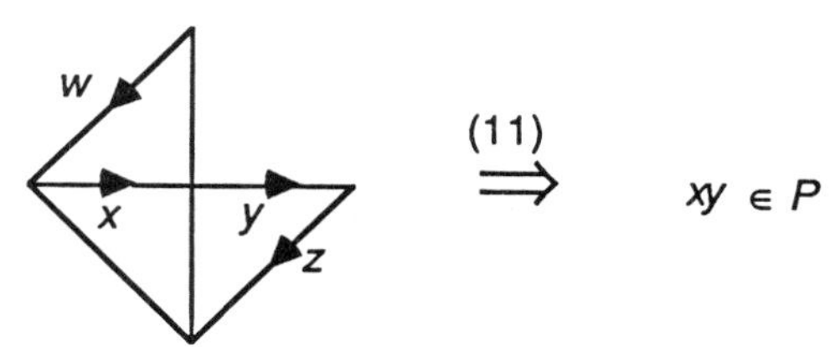

Case 2: $xyz \notin P$. Let $t = wx$. We have

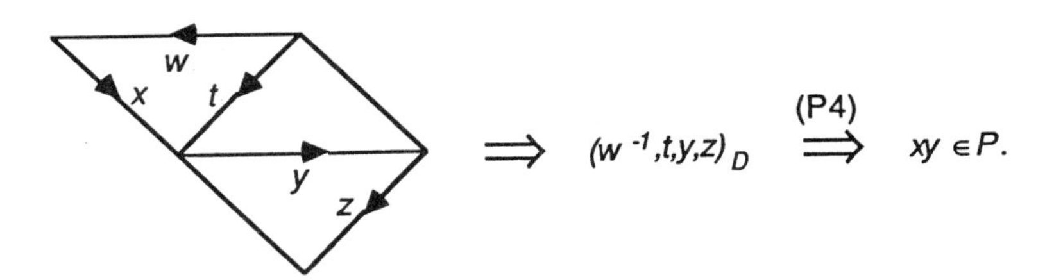

14. Corollary: 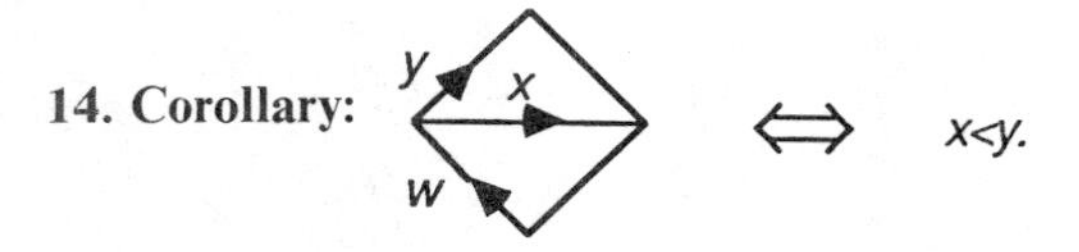

Proof: ($\Rightarrow$) By lemma 13, $zy \in P \Rightarrow zx \in P$ for all $z \in P$, so $x \lesssim y$. Since $wx \in P$ but $wy \notin P$, $x<y$.

($<=$) If $x<y$, then $x^{-1}y \in P$ by lemma 5. Since $y \not\lesssim x$, there exists $w \in P$ such that $wx \in P$, $wy \notin P$. □

Corollary 14 is the geometric interpretation of $x<y$ and is essential for the concept of *units*, which is introduced in section 3.

Recall from corollary 7 that $x \in P$ cyclic $\Rightarrow \{ x \lesssim x^{-1}$ or $x^{-1} \lesssim x \}$. In general, it is not true that $x \in P$ is cyclic $\Leftrightarrow$ $x \approx x^{-1}$ holds for all $x \in P$. Stallings has discovered some interesting counterexamples. However, we do have the following situation.

15. Definition: Let (P,D) and $\lesssim$ be as above. If every element of P is of finite height, and $\exists\ N \geq 0$ such that for all $x \in P$ the height of x is $\leq N$, then P is a pregroup with *finite height*. If N is minimal with respect to this property, then P is said to have *height* N. □

It is true that $x \in P$ cyclic $\Rightarrow$ $x \approx x^{-1}$ for all $x \in P$ provided P is of finite height. As we shall not need this result, we content ourselves with a weaker version following immediately from corollary 7. First we make an important definition.

16. Definition: Let (P,D) be a pregroup. Then $x \in P$ is *maximal* if $\nexists z \in P$ such that $x < z$. A *maximal vertex* V of P is an order tree vertex containing a maximal element of P. □

Clearly the maximal elements of P are the disjoint union of the maximal vertices of P.

17. Corollary: Let P be a pregroup, and let $x \in P$ be cyclic and maximal. Then $x \approx x^{-1}$. □

Every element in a pregroup with finite height is $\lesssim$ some maximal element. Pregroups of finite height will be our main objects of study. Every finite pregroup has finite height, and every pregroup whose order tree has a finite number of vertices has finite height. For any pregroup, V_1 is the unique vertex of height zero, but if P is of height N, then P could very well have an infinite number of vertices of height n for $1 \leq n \leq N$.

3. The subpregroup of units

Sections 3 and 4 develop the concepts that are necessary to describe the presentation of U(P) given in section 5. This presentation is in turn of fundamental importance in the proof of theorem A, stated in the introduction. In this section we define the *full subpregroup of units* of P and show how this leads to a decending sequence of supregroups of P.

Throughout this section, (P,D) will be a pregroup of finite height, and $\lesssim$ will be the tree ordering of P described in section 2. Let $M \subset P$ be the set of maximal elements of P with respect to the ordering $\lesssim$. Let $U = P - M$.

1. Definition: An element $x \in U = P - M$ is called a *unit* of P.

2. Lemma: Let $x \in P$. Then $x \in U \Leftrightarrow$

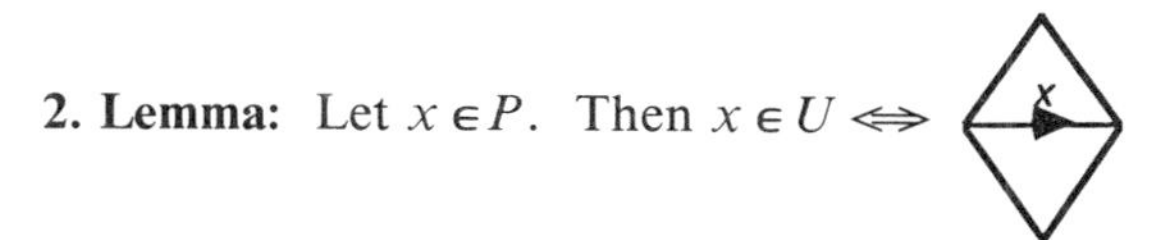

Proof: Use corollary 2.14. □

Notice that $U = \varnothing \Leftrightarrow P$ is a group.

3. Definition: Let $Q \subset P$ be a subset of P such that $1 \in Q$ and $x \in Q \Rightarrow x^{-1} \in Q$. Let $E = (Q \times Q) \cap D$. Suppose that $(x,y) \in E \Rightarrow xy \in Q$. Then (Q,E) is a *full* subpregroup of P. □

By verifying the pregroup axioms, we see that if Q is a full subpregroup of P, then Q is a subpregroup of P. By theorem 1.3, $Q \xrightarrow{\subset} P$ induces a group monomorphism $\mathrm{U}(Q) \xrightarrow{\phi} \mathrm{U}(P)$, such that $Q = \mathrm{im}\phi \cap P$.

4. Theorem: Suppose P is a pregroup that is not a group. Then the units U of P form a full subpregroup of P.

Proof: Since P is a pregroup that is not a group, corollary 1.9 implies there exists $x,y \in P$ such that $xy \notin P$. Thus

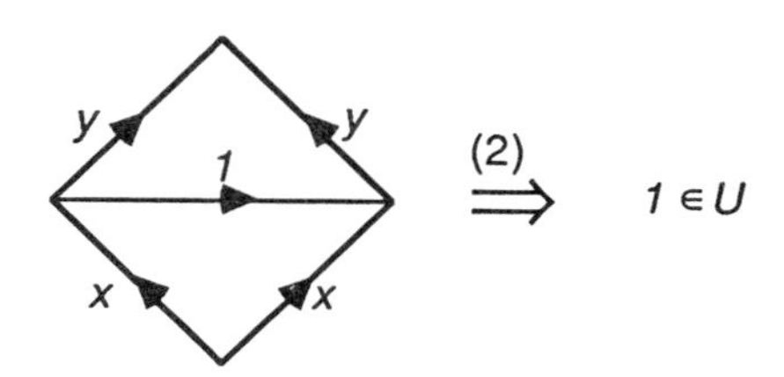

Clearly lemma 2 also implies U is closed under inversion. Now let $(x,y)\in(U\times U)\cap D$, so that $x,y\in U$ and $xy\in P$. We must show that $xy\in U$. Let $a,b,c,d,\alpha,\beta,\gamma,\delta\in P$ be such that

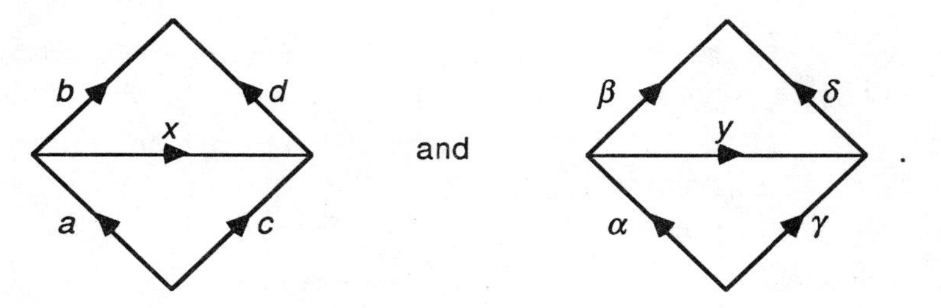

Let $z=xy\in P$.

Case 1: $cy\in P$. Let $f=cy$. Thus

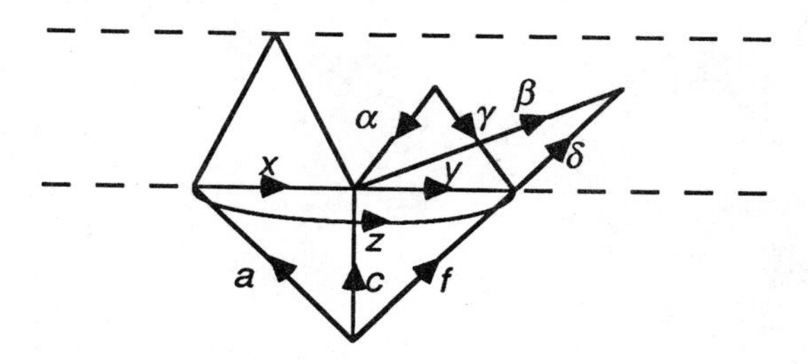

Case 1.1 $c\alpha^{-1}\in P$ and $c\beta\in P$. Then we have

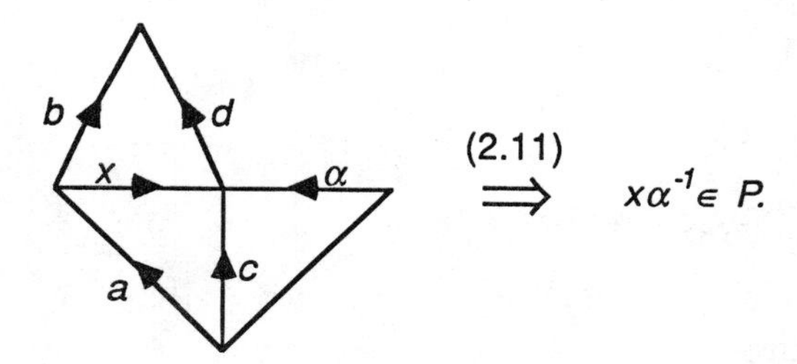

Similarly $x\beta\in P$. Let $r=x\alpha^{-1}$, $s=x\beta$. Then

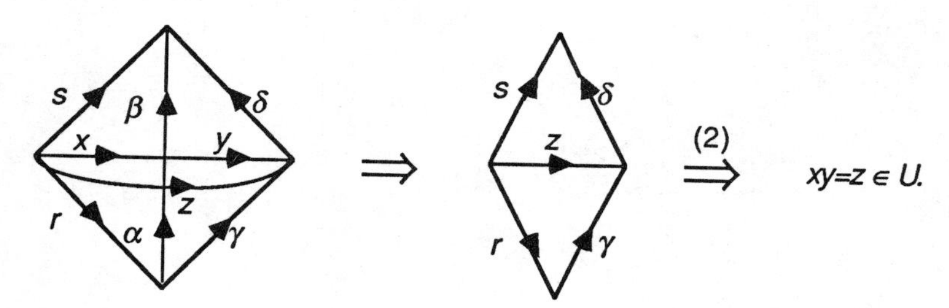

Case 1.2: $c\alpha^{-1} \notin P$ or $c\beta \notin P$. Then we have

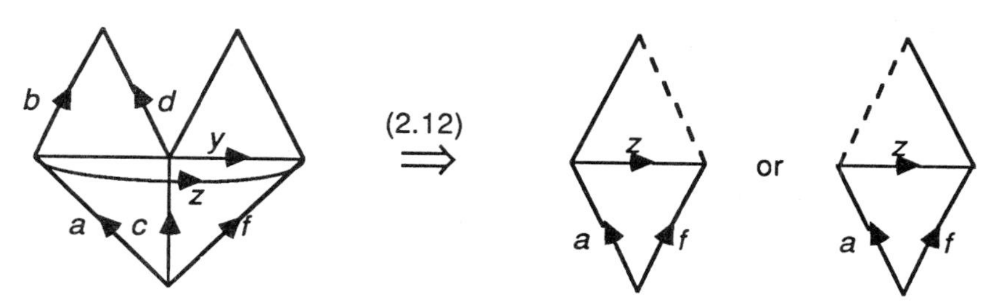

$\Rightarrow xy = z \in U$ by lemma 2.

Case 2: $cy \notin P$. Then we have

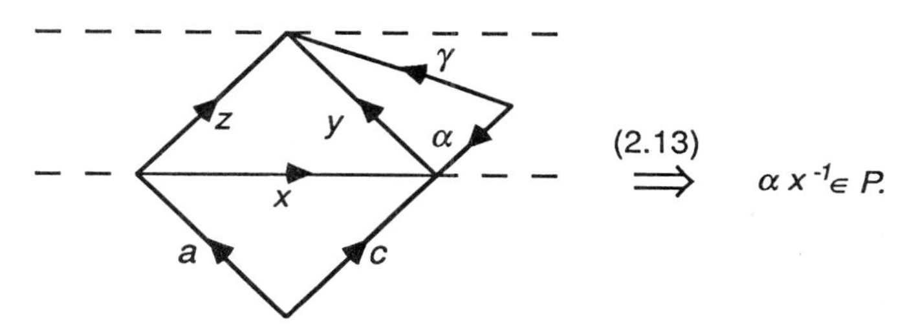

Apply the reasoning of case 1 to y^{-1} and x^{-1}. We see that $\alpha x^{-1} \in P \Rightarrow y^{-1}x^{-1} \in U$. Since U is closed under inversion, $xy \in U$ as desired. □

5. Corollary: The maximal elements of P are closed under inversion.

We deduce that the pregroups of finite height are built up by somehow adding on successive layers of maximal elements. However, caution must be taken because the order tree $(U, \lesssim_U)$ is in general not the restriction of $(P, \lesssim_P)$ to U. The following results clarify the situation.

6. Lemma: Suppose Q is a full subpregroup of P, and $x, y \in Q$. Then $x \lesssim_P y \Rightarrow x \lesssim_Q y$.

Proof: Suppose $x \lesssim_P y$. Let $z \in Q$, and suppose $z \bullet_Q y \in Q$, where $\bullet_Q$ is the multiplication for Q. Since Q is a subpregroup of P, $zy \in P$. Since $x \lesssim_P y$, $zx \in P$. Since Q is a full subpregroup of P, $z \bullet_Q x \in Q$. Thus $x \lesssim_Q y$. □

7. Example: (Stallings) Consider the minimal pregroup U for which the picture (z, x, y) is true. It has order tree

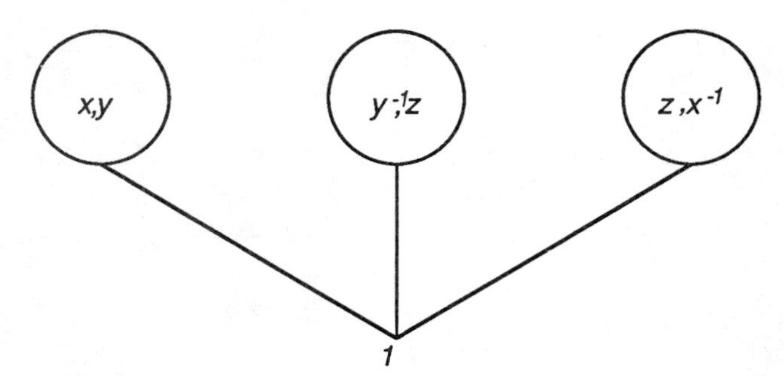

Now suspend the original picture, adding an extra flap above x.

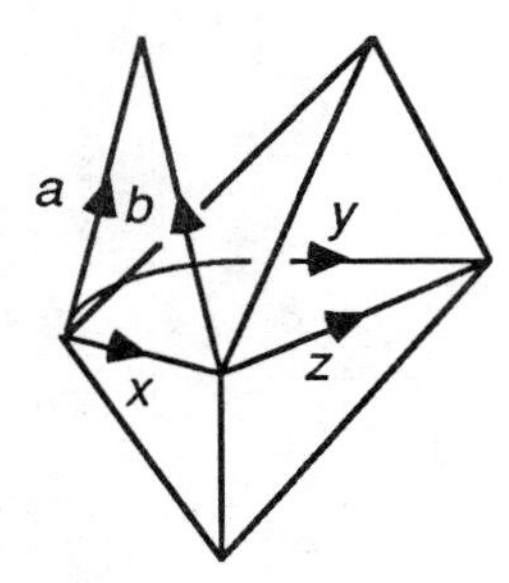

One may verify, (via computer in any event), that there is a minimal pregroup P for which this picture is true. Moreover, U is the set of units of P and the restriction of $\lesssim_P$ to U is

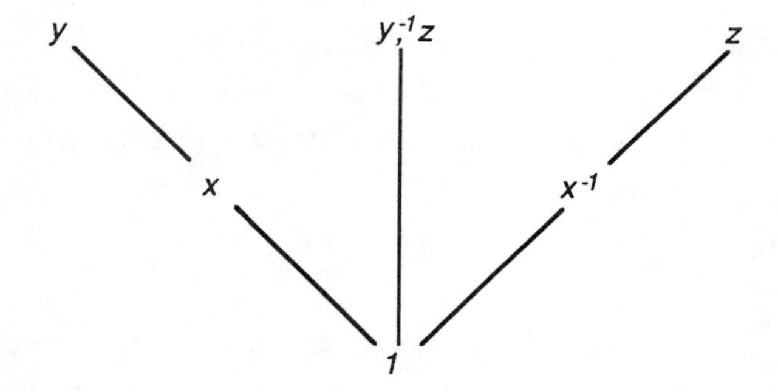

Thus $x \approx_U y$ but $x <_P y$. □

By lemma 6, every $\lesssim_P$-vertex is contained in a $\lesssim_U$-vertex, and so each $\lesssim_U$-vertex may be decomposed as a disjoint union of $\lesssim_P$-vertices. The above example shows that there can be more than one $\lesssim_P$-vertex in a given $\lesssim_U$-vertex.

However, we do have the following result.

8. Lemma: Suppose (Q,E) is a full subpregroup of P, and $x <_Q y$. Then $x <_P y$.

Proof: Let $x <_Q y$. Then by corollary 2.14 we have

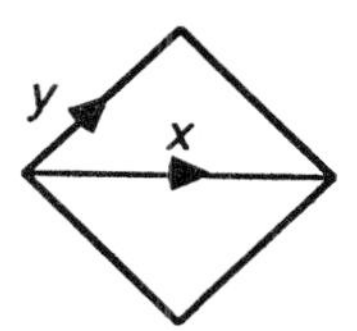

is a true statement with respect to the pregroup Q. Since Q is a full subpregroup of P, this statement is evidently true with respect to P. Thus $x <_P y$ by corollary 2.14. □

Stallings observed that lemma 8 implies that a full subpregroup of a pregroup of finite height is also of finite height.

Theorem 4 indicates that P can be filtered according to the following scheme. Given any pregroup P, let $U(P)$ be the units of P. By theorem 4, U takes pregroups which are not groups to pregroups. We use the operator notation $U^i(P)$ to denote i applications of U to P. Notice that for each $i \geq 0$, theorem 4 implies either $U^i(P)=\emptyset$ or $U^i(P)$ is a full subpregroup of P. If $U^m(P)$ is a group, then m is called the *depth* of P. In particular, groups are pregroups of depth 0.

9. Lemma: Let P be a pregroup of finite height $n \geq 1$. Then for some $m \leq n$, $U^{m+1}(P)=\emptyset$. Moreover, $U^m(P) \supset V_1$, the 0-vertex of P.

Proof: By definition, $P = M_0 \cup U^1(P)$, where M_0 is the set of maximal elements of $P = U^0(P)$. By theorem 4, $U^1(P)=\emptyset$ or is a pregroup in its own right. If $U^1(P) \neq \emptyset$, let M_1 be the set of maximal elements of $U^1(P)$. Hence $P = M_0 \cup M_1 \cup U^2(P)$. Keep repeating this argument so long as $U^i(P) \neq \emptyset$. Suppose $U^i(P) \neq \emptyset$ for some fixed integer i. Then $\exists\, m_0, m_1, \ldots, m_i$ such that $\forall\, j$, $0 \leq j \leq i$, $m_j \in M_j$, and $m_0 >_0 m_1 >_1 m_2 >_2 \cdots >_{i-2} m_{i-1} >_{i-1} m_i$. The "$j$" subscript on $>_j$ refers to the order tree for $U^j(P)$. Since each $U^j(P)$, $0 \leq j \leq i$, is a full subpregroup of P, lemma 8 implies $m_0 >_0 m_1 >_0 \cdots >_0 m_i$, so that $i \leq n$ by the definition of finite height.

Thus for some m, $0 \leq m \leq n$, $U^m(P) \neq \emptyset$, and $U^{m+1}(P)=\emptyset$. We claim that $U^m(P) \supset V_1$, the 0-vertex of P. Suppose $x \notin U^m(P)$. Then $x \in M_k$ for some $0 \leq k < m$, and evidently $1 <_k x$. By lemma 8, $1 <_0 x$, so that $x \notin V_1$. Thus $U^m(P) \supset V_1$. □

Incidentally, I do not know of any examples where $U^m(P)$ properly contains V_1. It seems unlikely that such a thing could happen. Lemma 9 may be paraphrased by saying that finite height implies finite depth.

10. Corollary: $U^m(P)$ is a subgroup of P, where m is as in lemma 9.

11. Corollary: Let $x \in P$ and let P be of height n. Then there exists a unique d, $0 \leq d \leq n$, such that $x \in U^d(P)$ and x is maximal in $U^d(P)$.

12. Definition: Let $x \in P$. Let d be such that x is maximal in $U^d(P)$. Then d is called the *depth* of x.

13. Remark: Let $x \in P$. If we consider a geodesic of minimal length in the order tree for P connecting V_x to a maximal vertex, then lemma 8 implies that the depth of x is bounded above by the length of this geodesic. However, this value may not be achieved. For instance, in example 7, the depth of x is 1 but the length of any geodesic from V_x to a maximal vertex of P is 2.

The following lemmas are of general importance.

14. Lemma: Let $x \in P$, and let d be the depth of x. Suppose x is cyclic. Then $x \approx_d x^{-1}$. Conversely, if $x \approx_d x^{-1}$, then x is cyclic.

Proof: Since x is cyclic, and $U^d(P)$ is a full subpregroup of P, x is a cyclic element of $U^d(P)$. Notice x is maximal in $U^d(P)$ by definition. Thus $x \approx_d x^{-1}$ by corollary 2.17. Conversely, if $x \approx_d x^{-1}$, then $x^2 \in P$. Corollary 1.10 implies that x is cyclic. □

15. Lemma: If $x, y, z \in P$, and $xy = z$, then either

(i) x, y, z are all units of P, or

(ii) x, y, z are all maximal, or

(iii) two elements of $\{ x, y, z \}$ are maximal and the other one is a unit.

Proof: Use unit × unit = unit (theorem 4). □

16. Lemma: If $x, y \in P$, x is maximal, and $xy \in P$, then $xy \lesssim x$.

Proof: Since $x^{-1}(xy) = y \in P$, $x \lesssim xy$ or $xy \lesssim x$ by lemma 2.6. Since x is maximal, $xy \lesssim x$. □

17. Lemma: Suppose x is maximal, $a \in P$, and ax, $x^{-1}a \in P$. Then a is cyclic.

Proof: If a is not cyclic, then 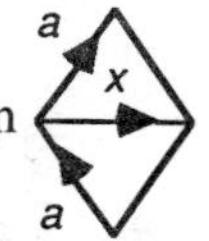

. By lemma 2, x is a unit, a contradiction. □

18. Lemma: Suppose $w, x, y \in P$, $wx, xy \in P$, and x is maximal. Then (w, x, y) associates.

Proof: If not then 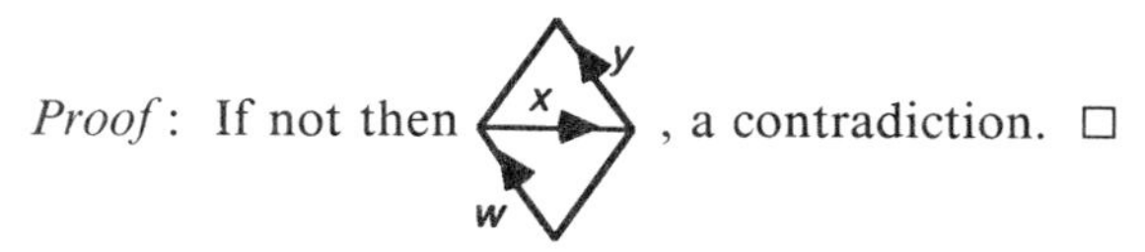

, a contradiction. □

19. Lemma: Let P be a pregroup, and let $x \in P$ be cyclic maximal. Suppose $a \in P$ is a unit, and $a<x$. Then (a^{-1},x,a) associates.

Proof: Since $a<x$, we know that x is *not* $\lesssim a$. Thus $\exists\, z \in P$ such that $za \in P$ but $zx \notin P$. Since x is cyclic, $a<x\approx x^{-1}$, so $xa \in P$. By axiom (P4), $(a^{-1},x,a,a^{-1}z^{-1})_D \Rightarrow \{ (a^{-1},x,a)$ associates or $xz^{-1} \in P \}$. But $xz^{-1} \in P \Rightarrow z^{-1} \lesssim x^{-1} \approx x \Rightarrow zx \in P$, a contradiction. Thus (a^{-1},x,a) associates. □

20. Lemma: Let $x \in P$ be cyclic maximal. Let $a \in U(P)$ and suppose $a<x$. Then $a^{-1}xa$ is cyclic maximal.

Proof: By lemma 19, (a^{-1},x^{-1},a) associates. By lemma 3.15, $a^{-1}xa$ and $a^{-1}x^{-1}a$ are maximal. But $(a^{-1}x^{-1})^{-1}(a^{-1}x)=x^2 \in P$, so $a^{-1}x \approx a^{-1}x^{-1}$. Hence $a^{-1}xa \approx a^{-1}x \approx a^{-1}x^{-1} \approx a^{-1}x^{-1}a$, so $a^{-1}xa$ is cyclic maximal. □

21. Lemma: Let (P,D) be a pregroup. Let $u,w \in U(P)$, and $x \in P$ be maximal. Let $w \in P$. Suppose $ux \in P$ and $(ux,w)_D$. Then (u,x,w) associates.

Proof: If not, then

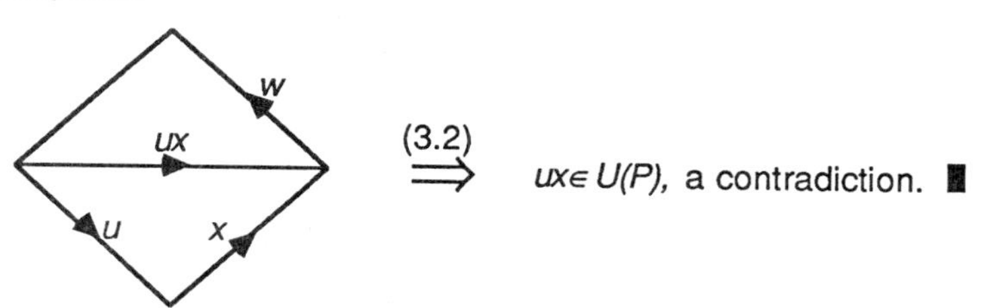

22. Lemma: Let (P,D) be a pregroup. Let $x,y \in P$ and suppose that y is maximal. Suppose $u \in U(P)$, and $uy \in P$. Furthermore, suppose $(x,uy)_D$. Then (x,u,y) associates.

Proof: If (x,u,y) does not associate, we have

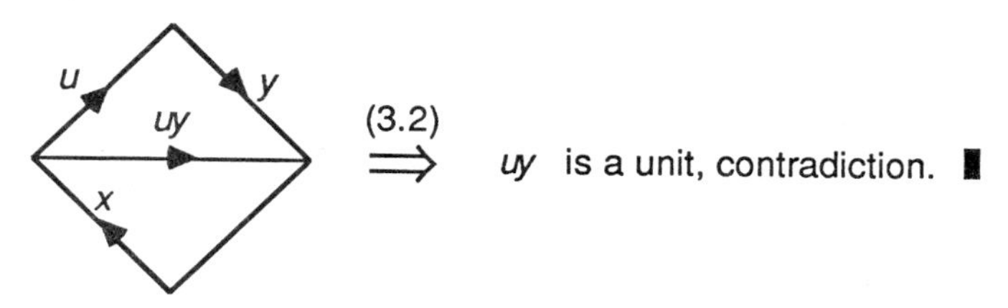

Part II
A presentation for the universal group of a pregroup of finite height

4. Pregroup actions and generating sets

In this section we continue the development started in section 3. We introduce the notion of a pregroup action, a straightforward generalization of a group action. Using this action, we define the *attaching group* and the *fundamental group* of a maximal vertex of P. We exploit the fact that the units act on the maximal vertices of P to find a nice generating set for $\mathrm{U}(P)$ inside of P. This generating set will consist of the units, certain fundamental groups, and *spanning sets,* defined below. These spanning sets were first considered by Stallings [1966] in the context of partial groups.

1. Definition: Let u be a unit of P, and let V be a maximal vertex of P such that $u^{-1}<x$ for some, (and hence all), $x\in V$. Then the set $uV\subset P$ is defined by the formula $uV=\{\ ux\in P \mid x\in V\ \}$.

2. Lemma: Let V be a maximal vertex of P. Suppose $x\in V$ and $u^{-1}<x$. Let V_{ux} be the maximal vertex of P containing ux. Then $V_{ux}=uV$.

Proof: Let $z\in V_{ux}$, so that $z\approx ux$. Hence $z^{-1}(ux)\in P$. By lemma 3.22, (z^{-1},u,x) associates. Hence $x^{-1}(u^{-1}z)\in P$, so $x\approx u^{-1}z$. Since $z=u(u^{-1}z)$, it follows that $V_{ux}\subset uV$. Now suppose that $z\in uV$, so that $z=uy$ for some $y\approx x$. Notice $(uy)^{-1}(ux)=y^{-1}x\in P$, so $z=uy\approx ux\in V_{ux}$. We conclude that $V_{ux}=uV$. □

This lemma indicates that the action of the units on the maximal vertices is well defined at least at the level of sets. To show this action reflects the pregroup structure of $U(P)$ we proceed as follows.

3. Definition: Let (P,D) be a pregroup. Let S be a set. Let $A\subset P\times S$ and fix a map $A\to S$, denoted by $(p,s)\to ps$. Then P *acts* on S with *action* A if $\forall\ x,y\in P$ and $s\in S$,

(A1) $(1,s)_A$ and $1s=s$

(A2) $(x,s)_A\Rightarrow(x^{-1},xs)_A$

(A3) $\{\ (y,s)_A$ and $(x,ys)_A\ \}\Rightarrow\{\ (x,y)_D,\ (xy,s)_A$, and $x(ys)=(xy)s\ \}$.

The notation $(x,s)_A$ means that $(x,s)\in A$.

Now fix a pregroup (P,D), a set S, and an action A of (P,D) on S.

4. Lemma: Given $s,t \in S$, set $s \sim t$ if there exists $x \in P$ such that $(x,s)_A$ and $t = xs$. Then $\sim$ is an equivalence relation on S.

Proof: Clearly (A1) implies $\sim$ is reflexive, (A2) implies $\sim$ is symmetric, and (A3) implies $\sim$ is transitive.

5. Definition: The $\sim$ classes of S are called *orbits* after the fashion of group actions.

6. Lemma: Let $s \in S$. Let $H_s = \{ x \in P \mid (x,s)_A \text{ and } xs = s \}$. Then H_s is a subgroup of P.

Proof: Clearly $1 \in H_s$. Suppose $x \in H_s$. Then by (A2), $(x^{-1},xs)_A$, so $(x^{-1},s)_A$ since $xs = s$. By (A3), $x^{-1}s = s$, hence H_s is closed under inversion. Suppose $x,y \in H_s$. Then $(x,s)_A \Rightarrow (x,ys)_A$, so by (A3) we have $(x,y)_D$, $(xy,s)_A$, and $(xy)s = x(ys) = xs = s$, so $xy \in H_s$. It follows that H_s is a subgroup of P.

7. Definition: H_s is called the *stabilizer* of s.

All the above lemmas and definitions apply in the case of arbitrary pregroups. Although it is not logically necessary to restrict ourselves to pregroups of finite height at every twist and turn, we shall do so for the rest of this section. Henceforth (P,D) is be a fixed pregroup of finite height greater than zero, and $U = U(P)$ is the pregroup of units of P. The boldface $\mathbf{V}$ is the set of maximal vertices of P. We say *action* instead of the more ponderous *pregroup action.* If $V \in \mathbf{V}$ and $u < x \in V$, we write $u < V$.

8. Lemma: There is an action A of U on $\mathbf{V}$ defined as follows. Let $(u,V) \in U \times \mathbf{V}$. Then $(u,V) \in A \Leftrightarrow u^{-1} < V$. In this case u acts on V by $V \to uV$.

Proof: By lemma 2, the action is well defined as a set map $A \to \mathbf{V}$. Evidently A satisfies axioms (A1) and (A2). Now suppose $V \in \mathbf{V}$, $x,y \in U$, $(y,V)_A$, and $(x,yV)_A$. Choose $z \in V$. Then $yz \in P$, and $yV = V_{yz}$, so $x(yz) \in P$. By lemma 3.22, $xy \in P$, hence $xy \in U$ since x and y are units. Thus $(xy)^{-1} < z \in V$, so $(xy,V)_A$, and $(xy)V = V_{(xy)z} = V_{x(yz)} = xV_{yz} = x(yV)$ as desired. □

Notation: Consider the action of U on $\mathbf{V}$. An orbit O of this action is a set $O = \{ V_\alpha \}$ of maximal vertices. In spite of this fact, it is more convenient to think of an orbit as a subset of maximal elements of P. Given an orbit O, let $\hat{O} = \{ x \in P \mid x \in V \text{ for some } V \in O \}$. By abuse of notation, we consider $\hat{O}$ to be an orbit, and we shall exclusively use this notion of orbit when symbolically notating such concepts as membership and inclusion. Consequently, the phrases "let V be a vertex of the orbit O" and "let $V \subset O$" are equivalent.

If an orbit O contains a cyclic element of P, then O is called a cyclic orbit. The boldface $\mathbf{O}$ is the set of orbits of P. □

We use this action to define the *attaching group* and the *fundamental group* of a maximal vertex of P.

9. Definition: Let V be a maximal vertex of P. The attaching group A_V of V is the stabilizer of V by the action of U on the maximal vertices of P. If $x \in V$, we also use the notation $A_x = A_V$, and refer the A_x as the attaching group of x.

10. Theorem: Let V be a maximal vertex of P. Let $G_V = A_V \cup \{ g \in V \mid g$ is cyclic $\}$. Then all of the following conclusions hold.

(i) $G_V \cap U(P) = A_V$,

(ii) $G_V = A_V \Leftrightarrow V$ is simple,

(iii) G_V is a subgroup of P,

(iv) if H is a subgroup of P, and $H \cap V \neq \emptyset$, then $H \subset G_V$.

Proof: (i) and (ii) are obvious. Clearly $1 \in G_V$ and G_V is closed under inversion. Let $x,y \in G_V$. If x and y are units, then $xy \in A_V \subset G_V$. If exactly one of x and y are units, we may assume x is the unit. Since y is cyclic by definition, $(xy)^{-1} \approx y^{-1} \approx y \approx xy$. Thus xy is cyclic, so $xy \in V$. Suppose both x and y are maximal. Then x and y are cyclic and $x,y \in V$. If xy is maximal, then $xy \approx x \approx y^{-1} \approx y^{-1}x^{-1}$, so $xy \in G_V$. If $xy \in U$, then $(xy)y^{-1} = x \in V$, so xy stabilizes V. thus $xy \in A_V \subset G_V$. Hence (iii) follows. Finally, suppose H is a subgroup of P and $x \in H \cap V$. If $y \in H$ is maximal, then $y^{-1}x \in H \Rightarrow y \approx x$, so $y \in V$. Therefore $y \in G_V$. If $y \in H$ is a unit, then $y^{-1}x \in H$ is maximal, hence $y^{-1}x \in V$. Thus $y(y^{-1}x) = x \in V$, so y stabilizes V. We conclude that $y \in A_V \subset G_V$. Thus $H \subset G_V$ as desired. □

11. Definition: Let V be a maximal vertex of P. Let $G_V = A_V \cup \{ x \in V \mid x$ is cyclic $\}$. Then G_V is called the fundamental group of V.

12. Lemma: Let V be a maximal vertex of P. Let G_V be the fundamental group of V. Let $m \in \mathbb{Z}$ be defined by $m = \min\{ \text{depth}(g) \mid g \in G_V \}$. Then there exists a unique $U^m(P)$-maximal vertex W such that $G_V \subset G_W$.

Proof: This follows from theorem 10.iv. □

13. Definition: Let V be a maximal vertex of P, with fundamental group G_V. We think of G_V as being *below* V. Let $m = \min\{ \text{depth}(g) \mid g \in G_V \}$. Then m is called the *depth of the fundamental group below* V. □

Notice the depth of the fundamental group below V is zero if and only if V is cyclic.

14a. Lemma: Let $x \in P$ be maximal. Let V be the maximal vertex containing x, and let $\tilde{V}$ be the maximal vertex containing x^{-1}. Then $a \in G_V \Leftrightarrow \{ a \in P$ and (x^{-1},a,x) associates $\}$. Moreover, $G_{\tilde{V}} = x^{-1} G_V x$.

Proof: If $a \in G_V$ and a is maximal, then $a \approx a^{-1} \approx x$, so that $(x^{-1},a,x)_D$. By lemma 3.18, (x^{-1},a,x) associates. If a is a unit, then $ax \in V$. Thus $x \approx ax$, so $x^{-1}(ax) \in P$. since $a<x$, $x^{-1}a \in P$. By pregroup axiom (P3), (x^{-1},a,x) associates.

Conversely, suppose that $a \in P$ and (x^{-1},a,x) associates. By lemma 3.17, a is cyclic. Since $x^{-1}a \in P$, $a \precsim x$. If a is maximal, then $a \approx x$, so $a \in G_V$. Now suppose that a is a unit. Then $ax \approx x$, so a stabilizes V. Thus $a \in A_V \cap G_V$, as desired.

It remains to prove that $G_{\tilde{V}}=x^{-1}G_V x$. Suppose $b \in G_{\tilde{V}}$. Then (x,b,x^{-1}) associates. Indeed, the first conclusion of lemma 14a has been demonstrated, and may be applied to $x^{-1} \in \tilde{V}$. Clearly (x^{-1},xbx^{-1},x) associates, so reasoning in the opposite direction yields $xbx^{-1} \in G_V$. We conclude that $G_{\tilde{V}} \subset x^{-1}G_V x$. Conversely, if $a \in G_V$, then (x^{-1},a,x) associates $\Rightarrow$ $x,x^{-1}ax,x^{-1})$ associates $\Rightarrow$ $x^{-1}ax \in G_{\tilde{V}}$. We conclude that $x^{-1}G_V x \subset G_{\tilde{V}}$, so $G_{\tilde{V}}=x^{-1}G_V x$, as desired. □

14.b Lemma: Let V be a maximal vertex, and let m be the depth of G_V below V. Suppose $u \in P$ is such that $u < V$ and depth$(u) > m$. Set $\tilde{V} = u^{-1}V$. Then $a \in G_V \Rightarrow \{ a \in P$ and (u^{-1},a,u) associates $\}$. Moreover, $G_{\tilde{V}}=u^{-1}G_V u$.

Proof: There exists $b \in G_V$ such that depth$(b)=m$ and $b \precsim V$. Hence b and u are comparable in P, and by lemma 3.6 they are comparable in $U^m(P)$. Since depth$(u) > m$, $u <_m b$. Let W be the $U^m(P)$ maximal vertex containing b. By lemma 12, $G_V \subset G_W$. Now suppose $a \in G_V$. If depth$(a)=m$, then $a \in W$. By lemma 3.20, (u^{-1},a,u) associates in $U^m(P)$, and therefore in P. Suppose depth$(a) > m$. Since $a \in G_W$, we have $a <_m W$. It follows that $u^{-1}a \in U^m(P)$. Moreover, $a^{-1} <_m W$, so $u^{-1}a^{-1} \in U^m(P)$ and therefore $au \in U^m(P)$. Notice $au <_m ab \in W$, so that $u^{-1}(au) \in U^m(P)$. It follows from axiom P3 that (u^{-1},a,u) associates in $U^m(P)$ and hence in P.

(Incidentally, the converse implication, analogous to that of lemma 14a, is false. Example 1 of section 6 yields a counter-example.)

Suppose $a \in G_V$. Then $u^{-1}a^{-1}u \precsim \tilde{V}$, so $u^{-1}au$ acts on $\tilde{V}$. We compute that $u^{-1}au\ \tilde{V} = V_{u^{-1}a} = V_{u^{-1}} = \tilde{V}$. Thus $u^{-1}G_V u \subset G_{\tilde{V}}$. Now suppose $b \in G_{\tilde{V}}$. As above, (u,b,u^{-1}) associates and $ubu^{-1} \subset G_V$. Hence $G_{\tilde{V}} \subset u^{-1}G_V u$, so $G_{\tilde{V}}=u^{-1}G_V u$ as desired. □

We are ready to exhibit the *generating set* for P promised above. The significance of the generating set will become clear in section 5. The first step is to define an equivalence relation $\simeq$ on **O** (cf. Stallings [1966]).

Henceforth let M be the set of maximal elements of P.

15. Lemma: Suppose O_1 and $O_2 \in \mathbf{O}$, and there exists $x \in M$ such that $x \in O_1$ and $x^{-1} \in O_2$. Let $V, W \in \mathbf{V}$ be such that $V \subset O_1$ and $W \subset O_2$. Then there exists $y \in M$ such that $y \in V$ and $y^{-1} \in W$.

Proof: By hypothesis there exists $u, v \in U$ such that $uV_x = V$ and $vV_{x^{-1}} = W$. Hence $ux \in V$ and $vx^{-1} \in W$. It follows from lemma 3.18 that (u, x, v^{-1}) associates, so that $uxv^{-1} \in M$. Set $y = uxv^{-1}$. Then $y = uxv^{-1} \approx ux \in V$ and $y^{-1} = vx^{-1}u^{-1} \approx vx^{-1} \in W$. □

16. Definition: Given O_1 and $O_2 \in \mathbf{O}$, say that $O_1 \simeq O_2 \Leftrightarrow \{ O_1 = O_2$ or there exists $x \in M$ such that $x \in O_1$ and $x^{-1} \in O_2 \}$.

17. Lemma: $\simeq$ is an equivalence relation on **O**.

Proof: Clearly $\simeq$ is reflexive and symmetric. Suppose $O_1 \simeq O_2$ and $O_2 \simeq O_3$. We may assume that $O_1 \neq O_2$ and $O_2 \neq O_3$, so that there exists $x, y \in M$ such that $x \in O_1$, x^{-1} and y are in O_2, and $y^{-1} \in O_3$. By lemma 15, we may arrange for $V_{x^{-1}} = V_y$. Thus $xy \in P$.

Case 1: $xy \in M$. Then $xy \approx x \in O_1$ and $y^{-1}x^{-1} \approx y^{-1} \in O_3$. Hence $O_1 \simeq O_3$.

Case 2: $xy \in U$. Then $xyV_{y^{-1}} = V_x$, so that $V_{y^{-1}}$ and V_x are in the same orbit. It follows that $O_1 = O_3$, so $O_1 \simeq O_3$ by definition. □

Let **E** be the set of $\simeq$-classes. (**E** stands for **E**quivalence). The convention established above for orbits also applies to $\simeq$-classes, so we consider the elements $E \in \mathbf{E}$ to be subsets of P.

Caution: Lemma 14b indicates that all the vertices of a cyclic orbit are cyclic. In general, it is *not* true that if $O_1 \simeq O_2$ and O_1 is cyclic, then so is O_2. Example 4 of section 5 exhibits this phenomenon.

18. Definition: For each $E \in \mathbf{E}$, choose a vertex $V(E) \in \mathbf{V}$ such that $V(E) \subset E$. The set $\mathbf{B} = \{ V(E) \in \mathbf{V} \mid E \in \mathbf{E} \}$ is called a *basepoint set for P*.

19. Definition: Let $E \in \mathbf{E}$. Let $V \in \mathbf{V}$ be such that $V \subset E$. From each orbit $O \subset E$ such that $V \cap O = \emptyset$, choose an element $x(O) \in M$ such that $x(O) \in O$ and $x(O)^{-1} \in V$. Denote this collection by span(V)= $\{ x(O) \mid O \subset E$ and $V \cap O = \emptyset \}$. We say that span(V) is a *spanning set for E based at V*.

20. Definition: Let **B** be a basepoint set for P. Let $\mathbf{G} = \{ (G_V, V) \mid V$ is a basepoint of $\mathbf{B} \}$. **G** is called the *fundamental group system* for **B**. The point of writing "(G_V, V)" instead of "G_V" is merely to distinguish fundamental groups of simple vertices which coincide as subsets of P. This nuance will be of no importance until section 6, where it will play a role in the proof of theorem A. We shall usually abuse notation by considering the elements of **G** as subgroups of P.

21. Definition: Let **B** be a basepoint set for P. For each $V \in \mathbf{B}$ choose a spanning set span(V) based at V. Then span(**B**)$= \{ x \in \text{span}(V) \mid V \in \mathbf{B} \}$ is called a *spanning set based at* **B**.

22. Definition: Let **B** be a basepoint set for P. Let **G** be the fundamental group system for **B**, and let span(**B**) be a spanning set based at **B**. Then the pair (**G**,span(**B**)) is called a *generating system for P,* and the set

$$GS = \{ x \in M \mid x \in G \text{ for some } G \in \mathbf{G} \text{ or } x \in \text{span}(\mathbf{B}) \}$$

is called the *associated generating set for P.* □

The following lemma is at the very heart of matters. The proofs of the minimality of the generating set, (theorem 24), and of the presentation for **U**(P), (theorem 5.3), depend on lemma 23 in a critical way. We will need the Cayley graph $\Gamma = \Gamma(\mathbf{U}(P),P)$ defined in section 2.

23. Lemma: Let P be a pregroup and $(p_1, \ldots, p_n)$ a P-word representing $1 \in \mathbf{U}(P)$. Suppose for some $i \in \mathbb{Z}$, $1 \leq i \leq n$, that $l_P(p_1 \cdots p_i) \geq 2$. Then there exists a unit $u \in U(P)$ and $j,k \in \mathbb{Z}$, $2 \leq j < k \leq n-1$, such that $p_j p_{j+1} \cdots p_k = u$ and $p_{k+1} \cdots p_1 \cdots p_{j-1} = u^{-1}$.

Proof: Let $m = \max\{ l_P(p_1 \cdots p_i) \in \mathbb{Z} \mid 1 \leq i \leq n \}$. Then $m > 1$ by hypothesis. Pick $j \in \mathbb{Z}$ such that $l_P(p_1 \cdots p_j) = m$ and $l_P(p_1 \cdots p_i) < m$ for $1 \leq i \leq j-1$. Pick $k \in \mathbb{Z}$ such that $l_P(p_1 \cdots p_k) = m-1$, and for all $i \in \mathbb{Z}$ such that $j < i < k$ we have $l_P(p_1 \cdots p_j \cdots p_i) = m$. Choose geodesics in Γ from 1 to $p_1 \cdots p_{j-1}$ and to $p_1 \cdots p_k$ as demonstrated in the following picture.

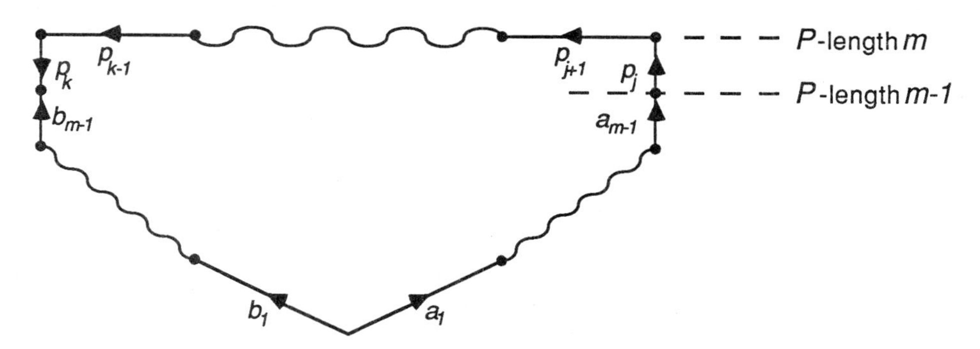

These geodesics are constructed by taking elementary reductions, (defined in section 1), of the P-words $(p_1, \ldots, p_{j-1})$ and $(p_{k+1}, \ldots, p_n)$. Hence $a_1 \cdots a_{m-1} = p_1 \cdots p_{j-1}$ and $p_{k+1} \cdots p_n = (b_1 \cdots b_{m-1})^{-1}$. Since $l_P(a_1 \cdots a_{m-1} p_k p_{k+1}) = m+1$ and $(a_1, \ldots, a_{m-1}, p_j)$ is P-reduced, theorem 1.4 implies $p_j p_{j+1} \in P$. We have

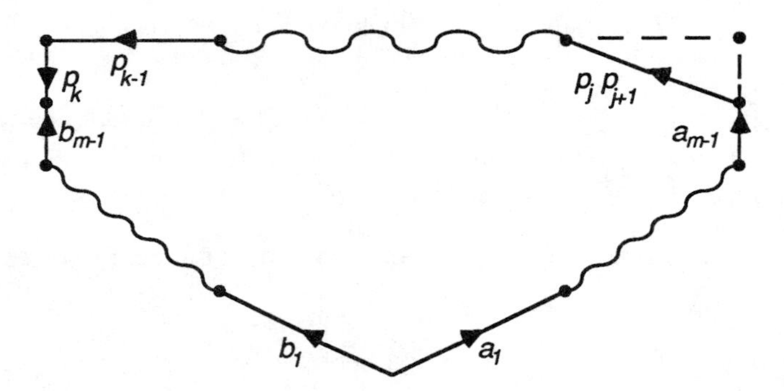

By repeating this step $k-j-1$ times we get

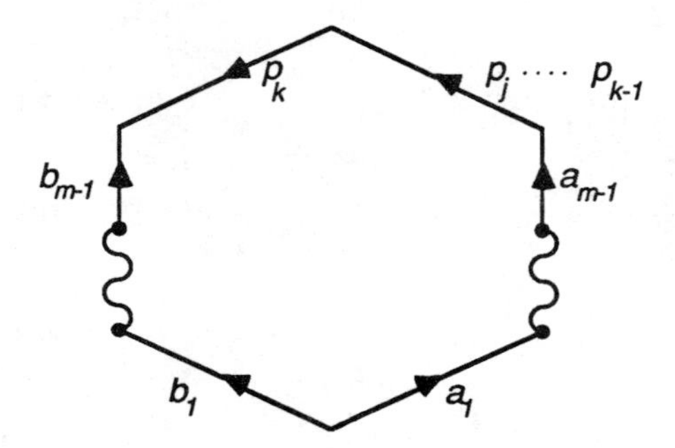

Hence $(a_1,...,a_{m-1},p_j \cdots p_{k-1})$ and $(b_1, \ldots, b_{m-1},p_k^{-1})$ are both P-reduced representing the same element of $\mathbf{U}(P)$. By theorem 1.4, these two words differ by an interleaved product. Thus we have

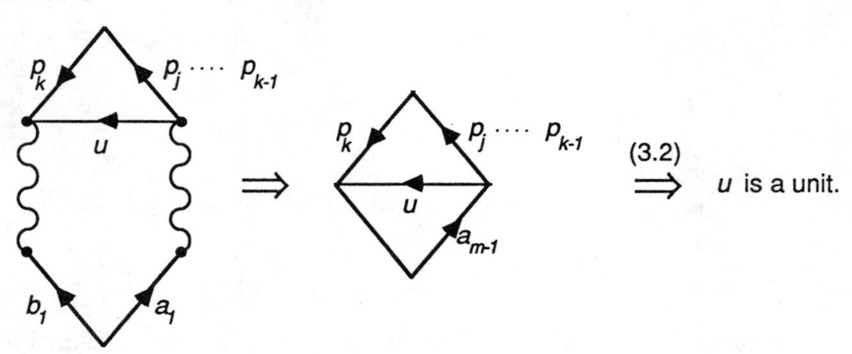

Moreover, $p_j \cdots p_k = u$ and $p_{k+1} \cdots p_n p_1 \cdots p_{j-1} = (b_1 \cdots b_{m-1})^{-1} a_1 \cdots a_{m-1} = u^{-1}$ as desired. □

24. Theorem: Let **B** be a basepoint set for P. Let (**G**,span(**B**)) be a generating system for P. Denote the associated generating set by GS. Let U be the set of units of P.

Then $U \cup$ GS generates $\mathbf{U}(P)$. Moreover, if $x \in$ GS is simple, and $G' = \mathrm{GS} - \{ x \}$, then $U \cup G'$ does *not* generate $\mathbf{U}(P)$.

Proof: Let S be the subgroup of $\mathbf{U}(P)$ generated by $U \cup$ GS. Ciearly it suffices to show that $P \subset S$. To this end, let $x \in P$. If $x \in U$, then evidently $x \in S$. So we assume that x is a maximal element of P. Let $O_x \in \mathbf{O}$ be such that $V_x \subset O_x$, and let $O_{x^{-1}} \in \mathbf{O}$ be such that $V_{x^{-1}} \subset O_{x^{-1}}$. Clearly $O_{x^{-1}} \simeq O_x$, so let $E \in \mathbf{E}$ be the $\simeq$-class containing O_x and $O_{x^{-1}}$. Let $V \in \mathbf{B}$ be such that $V \subset E$. Let O be the orbit containing V.

Case 1: $V_x \cap O \neq \emptyset$ and $V_{x^{-1}} \cap O \neq \emptyset$. Let span$(V) \subset V$ be the spanning set based at V such that span$(V) \subset$ span$(\mathbf{B})$. Let $y,z \in$ span(V) be such that $y \in O_x$ and $z \in O_{x^{-1}}$. Let $u,v \in U$ be such that $x \approx uy$ and $x^{-1} \approx vz$. Notice $y^{-1}u^{-1} \approx y^{-1} \approx z^{-1} \approx z^{-1}v^{-1}$ and $y^{-1}u^{-1}$ and $z^{-1}v^{-1}$ are in S. Thus we change notation, letting $y,z \in S \cap M$ be such that $x \approx y$, $x^{-1} \approx z$, $y^{-1} \approx z^{-1}$ and $V_{y^{-1}} = V \in \mathbf{B}$, the basepoint system for P. By lemma 3.18, (y^{-1},x,z) associates. Note $(y^{-1}xz)^{-1} = z^{-1}x^{-1}y \approx z^{-1} \approx y^{-1} \approx y^{-1}xz$, so $y^{-1}xz$ is cyclic maximal. Since $y^{-1}xz \in V_{y^{-1}} = V$, it follows that V is a cyclic basepoint of $\mathbf{B}$, so $y^{-1}xz \in G_V \subset \mathrm{GS} \subset S$. Hence $x \in S$.

Case 2: Exactly one of V_x and $V_{x^{-1}}$ is in O. We assume that $V_{x^{-1}} \subset O$. By hypothesis, $V_x \cap O = \emptyset$. As in case 1, let $V \in \mathbf{B}$ be such that $V \subset O$ and $z \in S \cap M$ such that $V_{z^{-1}} = V$ and $V_z = V_x$. Note that $x^{-1}z \in P$. Since $B_{x^{-1}} \subset O$, we may choose $u \in U$ such that $ux^{-1} \approx z^{-1}$. By lemma 3.22, (z^{-1},u,x^{-1}) associates. Since $x^{-1}z \in P$, $(z^{-1}u,x^{-1},z)$ associates by lemma 3.18. Notice that 3.18 also implies that (u,x^{-1},z) associates, so a final application of 3.18 yields (z^{-1},ux^{-1},z) associates. Thus $(z^{-1}ux^{-1}z)^{-1} \approx z^{-1} \approx z^{-1}ux^{-1}z$, so $z^{-1}ux^{-1}z$ is cyclic. As in case 1, $z^{-1}ux^{-1}z \in U$ or $z^{-1}ux^{-1}z \in G_V$, so we deduce that $x \in S$.

Case 3: Both V_x and $V_{x^{-1}}$ are in O. Let $V \in \mathbf{B}$ be such that $V \subset O$. Choose $u,v \in U$ such that $ux \in V$ and $vx^{-1} \in V$. By lemma 3.18, (v,x^{-1},u^{-1}) associates. Now $(vx^{-1}u^{-1})^{-1} = uxv^{-1} \approx ux \approx vx^{-1} \approx vx^{-1}u^{-1}$, so $vx^{-1}u^{-1}$ is cyclic. As in cases 1 and 2, we see that $x \in S$.

We conclude that $P \subset S$, or in other words, $U \cup$GS generates $\mathbf{U}(P)$. Now suppose $x \in$GS is simple. This can only happen if for some $V \in \mathbf{B}$, $x \in$ span$(V) \subset$ GS. Let $G' = \mathrm{GS} - \{ x \}$, and let S' be the subgroup of $\mathbf{U}(P)$ generated by $G' \cup U$. Let $\tilde{G} = \{ y \in M \mid y \in G' \text{ or } y^{-1} \in G' \}$. Thus every element in S' is represented by a word in $\tilde{G} \cup U$. We claim that if $y \in M \cap S'$, and O_y is the orbit containing y, then for some simple element of $z \in \tilde{G}$, $O_y = O_z$. By construction of the generating set,

$O_x \neq O_z$ for any $z \in \tilde{G}$, so $x \notin M \cap S'$. Evidently this implies that $x \notin S'$, as desired.

We proceed with the proof of the claim. Let $y \in M \cap S'$. Let $(y_1, \ldots, y_n)$ be a P-word of minimum length representing y such that each $y_i \in G' \cup U$.

Subclaim: For each $i \in \mathbb{Z}$ such that $1 \leq i \leq n$, $y_1 \cdots y_i \in P$.

Proof: Suppose not. Apply lemma 23 to the P-word $(y_1, \ldots, y_n, y^{-1})$. Thus for some $1 < j < k \leq n$, $y_{k+1} \cdots y_n \cdots y^{-1} y_1 \cdots y_{j-1} = u^{-1}$, where $u \in U$. Hence $y = y_1 \cdots y_{j-1} u y_{k+1} \cdots y_n$. This contradicts minimality of n. □

By "conjugating" the proof of the subclaim, we can show that for $i, j \in \mathbb{Z}$, such that $1 \leq i < j \leq n$, we must have $y_i \cdots y_j \in P$. We now show that $O_y = O_z$ for some z in $\tilde{G}$ by considering the three possibilities $n=1$, $n=2$, or $n > 2$.

Case 1: $n=1$. Clearly $y \in \tilde{G}$, so setting $y=z$, we see $O_y = O_z$ for some $z \in \tilde{G}$.

Case 2: $n=2$. Thus $y = y_1 y_2$, so by minimality of n, $y \in M$. If $y_1 \in U$, then $y_2 \in M$ and $O_y = O_{y_2}$. If $y_1 \in M$, then $y \approx y_1$, so evidently $O_y = O_{y_1}$.

Case 3: $n > 2$. Let $w = y_3 \cdots y_n$. Then $y_1 y_2 \in P$ and $w \in P$ by the claim. Since $y = (y_1 y_2) w$, $y \approx y_1 y_2$. As in case 2, $O_y = O_{y_2}$ or O_{y_1}, depending on whether $y_1 \in U$ or in M. □

5. A presentation for the universal group of a pregroup of finite height

In this section we give a presentation for $U(P)$ in terms of a generating system for P. The presentation relies heavily on the notions of the fundamental groups and the generating set of a pregroup. These ideas were developed in sections 3 and 4. Before stating the result we discuss coproducts and an alternative characterization of the attaching group of a maximal vertex. After the proof of the main result, we provide an example that illustrates the the presentation, and also provides motivation for the proof of theorem A in section 6. Specifically, in example 4 of this section we obtain a graph of groups from a generating system for a pregroup according to the algorithm set forth precisely in the proof of theorem A. (Insofar as our example is of height one, we avoid the major technical difficulty dealt with in section 6). We end this section with a short proof that the pregroups of finite height with no non-trivial cyclic elements have free universal group.

In view of the adjointness relationship between groups and pregroups, coproducts exist in the category of pregroups. In fact, the coproduct of two pregroups is just their disjoint union with identity elements identified. If P and Q are pregroups, then $U(P \oplus Q)=U(P) * U(Q)$. Given any set X, the free pregroup on X, $FP(X)$, is constructed as follows. Let X^{-1} be a disjoint copy of X, and set $FP(X)=X \cup X^{-1} \cup 1$. The partial multiplication for $FP(X)$ is such that reduced $FP(X)$-words correspond with freely reduced words in the usual sense. In particular, $U(FP(X))$ is the free group on X.

1. Recall: Let P be a pregroup and let V be a maximal vertex of P. Then the attaching group of V is the stabilizer of V in $U(P)$, denoted A_V. The fundamental group G_V of V is the union of A_V and the cyclic elements of V, if any.

2. Remarks: (1) Suppose V is a maximal vertex and $x \in V$. Lemma 4.14a implies that $A_V = \{ a \in U(P) \mid (x^{-1},a,x) \text{ associates } \}$. Hence A_V is also denoted A_x. We also denote G_V by G_x.

(2) Suppose x and V are as above. Then lemma 4.14a implies that $x^{-1}A_x x \subset G_{x^{-1}}$. However, it may not be true that that $A_{x^{-1}}=x^{-1}A_x x$. Example 4 of this section contains an example in which a conjugate $x^{-1}ax$ of an element of the attaching group A_x is maximal, and therefore certainly not contained in $A_{x^{-1}}$.

We are now ready to state our main result.

3. Theorem: Let (P,D) be a pregroup of finite height. Let **B** be a basepoint system for P. Let (**G**,span(**B**)) be a generating system for P. Index **G** by $\mathbf{G}=\{ G_\alpha \}$. Let X=span(**B**). Let U be the set of units of P. Let $Q=FP(X)\oplus U\oplus_\alpha G_\alpha$. Denote multiplication in Q by $x\bullet y$. Then there is a pregroup morphism $Q\to P$ which is an inclusion map on each component of Q, and the induced map $\phi:\mathrm{U}(Q)\to\mathrm{U}(P)$ is an epimorphism with kernel normally generated by the set R, where

$$R=\{ a\bullet b^{-1}\in Q \mid a,b\in Q,\ \phi(a)=\phi(b)\in U \}$$
$$\cup\{ x^{-1}\bullet a\bullet x\bullet b^{-1}\in Q \mid \phi(x)\in X,\quad \phi(a)\in A_{\phi(x)},\quad \phi(b)\in G_{\phi(x^{-1})}, \text{and } \phi(x^{-1}\bullet a\bullet x)=\phi(b) \}$$

Proof: It is convenient to identify certain elements of Q and P in a systematic way. Schematically,

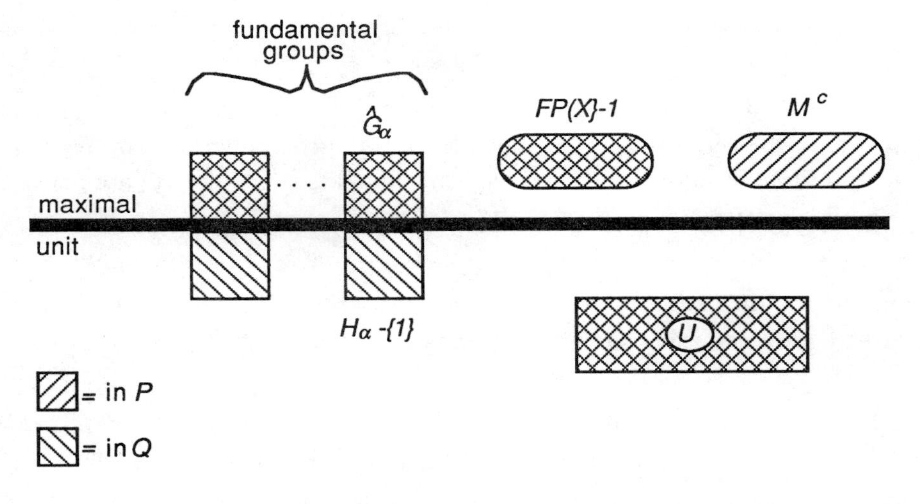

Let M be the set of maximal elements of P. For each G_α in the generating set, let $\hat{G}_\alpha=G_\alpha\cap M$. Then by definition, if $\alpha\neq\beta$, then $\hat{G}_\alpha\cap\hat{G}_\beta=\emptyset$. Thus the disjoint union $\dot{\bigcup}_\alpha\hat{G}_\alpha\dot{\cup}U$ may be identified with $\bigcup_\alpha\hat{G}_\alpha\cup U$ and considered to be a subset of P. Similarly, we identify $FP(X)$ with its image $\phi(FP(x))\subset P$. Letting $M^c=M-\bigcup_\alpha\hat{G}_\alpha-FP(X)$, we may assume that

$$P=(\bigcup_\alpha\hat{G}_\alpha\cup U)\cup M^c\cup FP(X) \text{ and}$$

$$Q=(\bigcup_\alpha\hat{G}_\alpha\cup U\cup FP(X))\dot{\bigcup_\alpha}(G_\alpha-\hat{G}_\alpha).$$

We also demand that $P \cap Q = \bigcup_{\alpha} \hat{G}_\alpha \cup U \cup FP(X)$. To this end we construct disjoint copies H_α of the groups $G_\alpha - \hat{G}_\alpha$, with identity elements identified. Each H_α is in $Q-P$. This inclusion is such that $\phi: \mathbf{U}(Q) \to \mathbf{U}(P)$ restricted to $H_\alpha \cup \hat{G}_\alpha$ and "co-restricted" to $G_\alpha \subset P$ is a group isomorphism. The Venn diagram may help clarify this point. Notice that the "original" $G_\alpha \subset U \cup \hat{G}_\alpha$ is in $Q \cap P$. The set $\phi^{-1}(U)$ is denoted U_Q, and $U_Q = U \bigcup_{\alpha} H_\alpha$.

Caution: Do not confuse U_Q with $U(Q)$, the set of units of Q. $U(Q)$ is either empty or trivial, depending on whether Q is or is not a group.

We have labored so carefully to set up this notation for the following reason. If $x,y \in \hat{G}_\alpha$ for some fundamental group G_α, we may infer that $x \bullet y \in \hat{G}_\alpha \cup H_\alpha \subset Q$. In this case $l_Q(x \bullet y) = l_P(xy) = 1$. On the other hand, suppose x and y are in U_Q. Evidently $\phi(x)$ and $\phi(y)$ are in U, and if $\phi(x)\phi(y) \notin U$, then $x \bullet y \notin Q$. However, suppose $\phi(x)\phi(y) \in P$. Then $x \bullet y$ may or may not be in Q. In so far as our entire argument is based on reducing the Q-length of various words in Q, this point should be noted with care.

We establish the following proposition:

Proposition 3a: Let $q \in \ker \phi$. Let $K = \{ y^{-1} \bullet r \bullet y \mid y \in \mathbf{U}(Q), r \in R \}$. Then

(i) $q \in K$, or $q^{-1} \in K$, or

(ii) for some $j \in \mathbb{Z}$, q or $q^{-1} = r_1 \bullet \cdots \bullet r_j$, such that

(a) each $r_i \in \ker \phi$, and

(b) each $r_i \in K$ or else $l_Q(y^{-1} \bullet r_i \bullet y) < l_Q(q)$ for some $y \in \mathbf{U}(Q)$. □

(Granting this proposition, induction on the Q-length of $q \in \ker \phi$ shows that q is in the subgroup normally generated by R in $\mathbf{U}(Q)$.)

Proof: (of proposition 3a). Now given $q \in \ker \phi$, choose $W = (q_1, \ldots, q_n)$ to be Q-reduced representing q. We may assume W is cyclically Q-reduced, for otherwise $l_Q(q_1^{-1} \bullet q \bullet q_1) < l_Q(q)$, yielding the result. Let $\phi(W) = (p_1, \ldots, p_n) = (\phi(q_1), \ldots, \phi(q_n))$ so that $\phi(W)$ is a P-word representing $1 \in P$.

Case 1: For some $i \in \mathbb{Z}$, $1 \le i \le n$, we have $l_P(p_1 \cdots p_i) > 1$. By lemma 4.23, we have $j,k \in \mathbb{Z}$, $2 \le j < k \le n-1$ and $u \in U$ such that $u = p_j \cdots p_k$ and $u^{-1} = p_{k+1} \cdots p_1 \cdots p_{j-1}$. Now set $r_1 = q_j \bullet \cdots \bullet q_k \bullet u^{-1}$ and $r_2 = u \bullet q_{k+1} \bullet \cdots \bullet q_1 \bullet \cdots \bullet q_{j-1}$. Then r_1, $r_2 \in \ker \phi$ and have smaller Q-length then q. Since q is a product of conjugates of r_1 and r_2, we are done. □

Henceforth, an argument of this type will be referred to as a *cut along u*. Formally, if there exists $s \in Q \cap P$ and $j,k \in \mathbb{Z}$, $2 \le j < k \le n-1$ such that $p_j \cdots p_k = s$

and $p_{j+1} \cdots p_1 p_{k-1} = s^{-1}$, then we may cut along s, establishing the proposition. The rest of the proof is a search for such cuts in various situations.

By case 1 and conjugation, we assume that $l_P(p_i \cdots p_k) \leq 1$ for any cyclic segment of $\phi(W)$.

Case 2: $q_1, q_2 \in U_Q$. Then case 1 implies that $p_1 p_2 \in P$ and hence $p_1 p_2 = u \in U$. If $n=2$, then $q_1 \bullet q_2 \in R$ by the definition of R. If $n > 2$, notice $q_1 \bullet p_1^{-1}$ and $q_2 \bullet p_2^{-1}$ are both in R. We are considering p_1 and p_2 as elements of Q in the manner described above. Set $r_1 = q_1 \bullet p_1^{-1}$, $r_2 = u \bullet (p_2^{-1} \bullet q_2) \bullet u^{-1}$, and $r_3 = u \bullet q_3 \bullet \cdots \bullet q_n$, so that $r_1, r_2, r_3 \in \ker \phi$. We have arranged things so that $p_1^{-1} \bullet u = p_2$, so evidently $r_1 \bullet r_2 = q_1 \bullet q_2 \bullet u^{-1}$. Hence $r_1 \bullet r_2 \bullet r_3 = q$. Since $l_Q(r_3) \leq n-1$, we have satisfied the proposition. □

Henceforth we assume that no two cyclically adjacent p_i, p_{i+1} are both units.

Case 3: For some G_α, G_β in the generating set, $q_1 \in \hat{G}_\alpha \cup H_\alpha$, $q_2 \in Q_U$, and $q_3 \in \hat{G}_\beta \cup H_\beta$. By case 2, we may assume $q_1 \in \hat{G}_\alpha$ and $q_3 \in \hat{G}_\beta$, so q_1 and $q_3 \in Q \cap P$. For the sake of vividness, let $a = q_1$, $b = q_3$, and set $u = \phi(q_2) \in U$. By case 1, (a, u, b) associates in P. Evidently a and b are cyclic maximal elements of P.

> Case 3.1: $l_Q(q) = n = 3$. Then $aub = 1$, so $b = u^{-1} a^{-1}$. Hence $b \approx b^{-1} = au \approx a$ in P, so both a and b are in $\hat{G}_\alpha$. This cannot happen since q is supposed to be cyclically Q-reduced.
>
> Case 3.2: n>3. If $au \in P$ is cyclic, then $au \in \hat{G}_\alpha$ since $a \approx au$. Setting $s = au \in Q \cap P$, $j=2$, and $k=3 < n$, we see that we can cut along s in the manner outlined above. So suppose au is simple. Notice $(au)b \in P$, so $b \approx (au)^{-1}$. By definition 4.16, $V_a = V_{au}$ and $V_b = V_{(au)^{-1}}$ are in the same $\approx$-class. By construction of the generating set, definition 4.22, $G_\alpha = G_\beta$, and therefore $V_a = V_b$. Since au is simple $V_a \neq V_b$. This contradiction shows that the assumption that au is simple is an absurdity, and therefore brings this case to a happy conclusion. □

By case 3, we may assume that W contains a simple maximal element $x \in FP(X) \subset Q \cap P$, and hence we assume that $q_1 = x$. By minimality of the generating set, as expressed in theorem 4.24, for some i between 2 and n, $q_i = x$ or $q_i = x^{-1}$.

Case 4: $q_1 = x$, and $q_i = x$ for some $2 \leq i \leq n$.

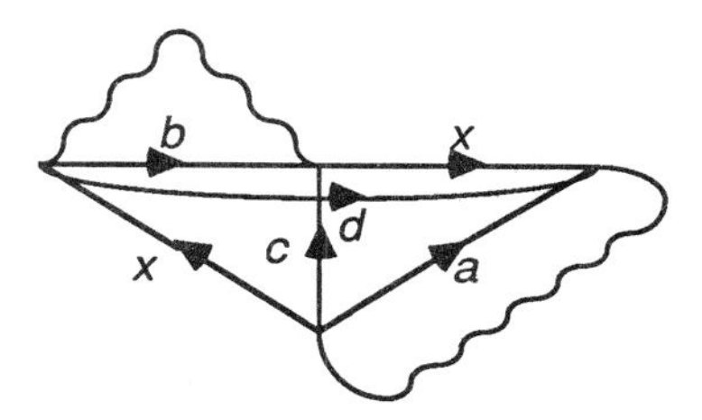

Since x is simple, we deduce that $2 < i < n$. Set $a = (p_{i+1} \cdots p_n)^{-1}$ and $b = p_2 \cdots p_{i-1}$. Let $c = xb$. Then $cx \in P$ and $c^{-1}x \in P$, so $c^{-1} \lesssim x$ and $c \lesssim x$, so c is cyclic. If c is a unit, we may cut along c. So assume c is maximal. Thus $c \approx x$. Similarly, we may assume $d = bx$ is cyclic maximal and $d \approx x^{-1}$. By construction of the generating set, either G_c or G_d is in the fundamental group system $\mathbf{G}$, so we can cut along c or along d. □

So for our final case, we may assume that W contains exactly two occurrences of some simple maximal element $x \in FP(X)$ which differ in sign.

Case 5: $p_1 = x$, $p_i = x^{-1}$ for some $i \in \mathbb{Z}$, $2 < i < n$, and $p_j \neq x$ or x^{-1} for $j \neq 1$ or i. By conjugating q if necessary, we may assume that $x \in X$. Thus definitions 4.19,4.20, and 4.22 together imply that $G_{x^{-1}} \in \mathbf{G}$, the fundamental group system. Therefore, the elements of $G_{x^{-1}}$ are in $Q \cap P$. It is equally important to observe that no maximal element of G_x is in Q.

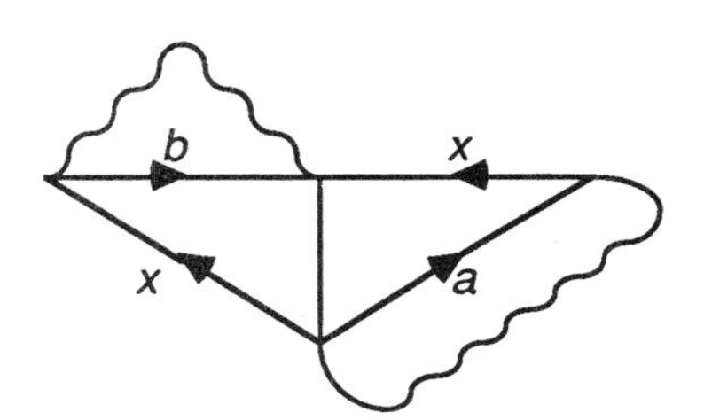

Define a and b as in case 4. Lemma 3.17 implies b is cyclic. Since $xb \in P$, $b \lesssim x^{-1}$, and therefore $b \in G_{x^{-1}} \subset Q \cap P$. Hence if $i \neq 3$, we may cut along b. So suppose $i = 3$. Let us first deal with some easy cases.

Case: a is a unit, and $n = 4$: then $a \in A_x$, and therefore $x^{-1} \bullet q_4^{-1} \bullet x \bullet q_2^{-1} \in R$. Since q^{-1} is a conjugate of this element, we are done.

Case: a is a unit, and $n > 4$: then we may cut along a.

So suppose a is maximal. By lemma 3.17, a is cyclic, and clearly $a \approx x$. Thus a is a maximal element of G_x, and therefore $a \notin Q$. Thus $n > 4$ and we set $u = p_4$.

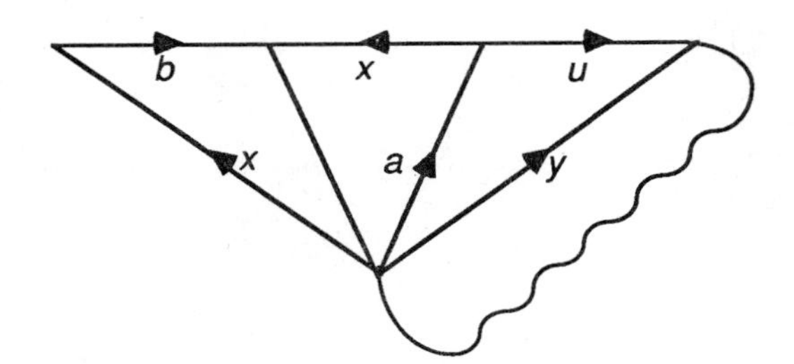

Then $u \lesssim x$, so if u is maximal, $u \in Q \cap V_x = x$, and therefore $u = x$. But this contradicts the fact that W is Q-reduced. Thus u is a unit.

Case 5.1: Suppose $y = au \in Q$. Then since $y \approx a$, $y = x$. Let $r_1 = q_2 \bullet x^{-1} \bullet q_4$, $r_2 = q_5 \bullet \cdots \bullet q_n \bullet x$. Then $r_1, r_2 \in \ker \phi$, and $q = x \bullet r_1 \bullet r_2 \bullet x^{-1}$, so we are done. □

Hence we may assume that $y \notin Q$. Then $n > 5$, so let $c = p_5$, $z = yc$.

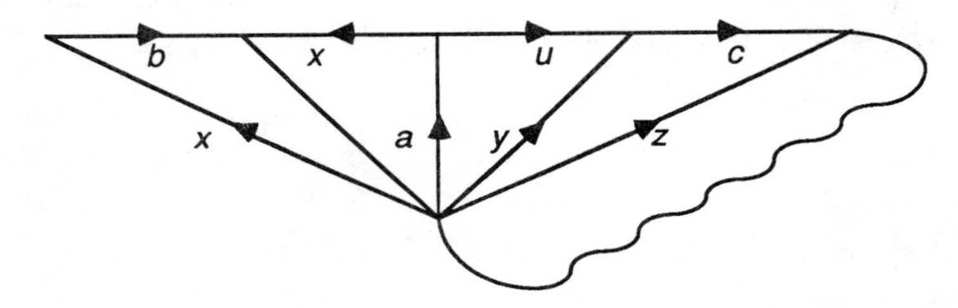

By case 2, c is maximal.

Case 5.2: Suppose z is a unit. Notice that $c = u^{-1}xb^{-1}x^{-1}z$. If c were simple, then minimality of the generating set implies $c = x$ or x^{-1}. Since there are only two occurrences of x in W by assumption, we deduce that c is cyclic. Since $c \in Q$, it follows that the fundamental group G_c containing c is in **G**. Since $y = zc^{-1}$, y is maximal. Moreover, $V_y = zV_c$, so V_y and V_c are in the same orbit. Since $x^{-1}y \in P$, $x \in V_y$. Hence x and G_c are in the same orbit, so $x \in V_c$ by construction of the generating set. Recall that a is cyclic maximal and $a \notin Q$. In particular, $V_a \neq V_c$, because $G_c \in \mathbf{G}$. Nevertheless, $x^{-1}a \in P$, so $x \approx a$, so $V_a = V_x = V_c$, a contradiction. Therefore z must be maximal.

Case 5.3: Suppose z is maximal. We may assume y is maximal, for otherwise we could cut along y.

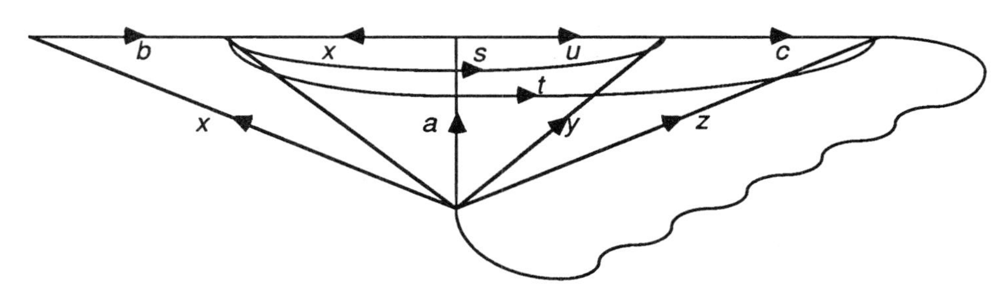

Now $y \approx z \approx x$ and $y^{-1} \approx c$, $z^{-1} \approx c^{-1}$. Thus V_c, $V_{c^{-1}}$, V_x, and $V_{x^{-1}}$ are all in the same $\approx$-class. By construction of the generating set, either $G_c \in \mathbf{G}$ or $G_{c^{-1}} \in \mathbf{G}$. Hence either $c \approx x^{-1}$ or $c^{-1} \approx x^{-1}$. Let $s = x^{-1}u \in P$. Notice $s \lesssim x^{-1}$ and $s^{-1} \lesssim c$, so if $c \approx x^{-1}$ then $s \in G_{x^{-1}} \cup U$, and we can cut along s. Similarly, let $t = x^{-1}uc \in P$. Then $t \lesssim x^{-1}$ and $t^{-1} \lesssim c^{-1}$, so if $c^{-1} \approx x^{-1}$, then $t \in G_{x^{-1}} \cup U$ and we can cut along t. □

As stated, this result is fairly indigestible. But it is clear that the presentation *resembles very strongly* a presentation for a graph of groups. In the next example we explore this connection in a somewhat intuitive manner. This example should be kept in mind when we introduce the more formal notation of section 6.

4. Example: Consider the pregroup P with multiplication table

	g	x^{-1}	gx^{-1}	xgx^{-1}	z	$xgx^{-1}z$	x	xg	z^{-1}	$z^{-1}xgx^{-1}$
g	1	gx^{-1}	x^{-1}							
x	xg	1	xgx^{-1}							
xg	x	xgx^{-1}	1							
xgx^{-1}				1	$xgx^{-1}z$	z	xg	x		
z^{-1}				$z^{-1}xgx^{-1}$	1					
$z^{-1}xgx^{-1}$				z^{-1}		1				
x^{-1}				gx^{-1}			1	g		
gx^{-1}				x^{-1}			g	1		
z									1	xgx^{-1}
$xgx^{-1}z$									xgx^{-1}	1

The elements g and xgx^{-1} are of order 2. All the other non-identity elements are simple. I have verified by computer that this really is a pregroup. From the multiplication table we deduce the following order tree.

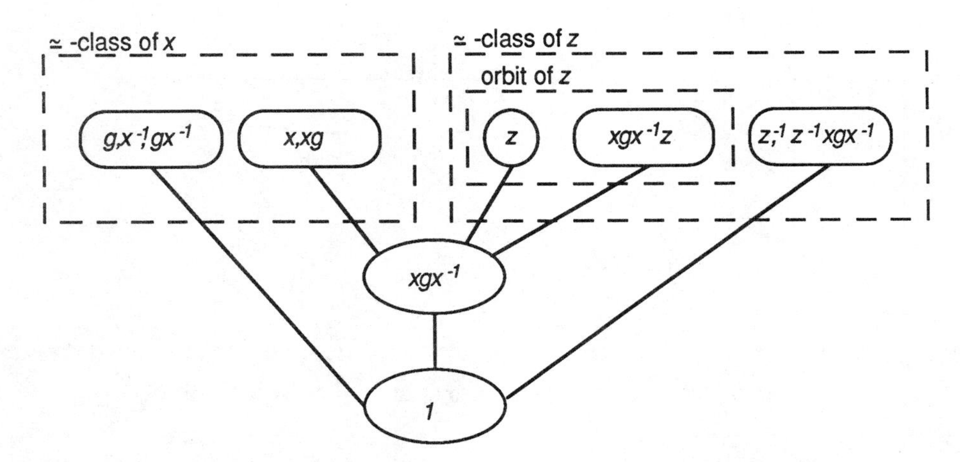

This example possesses many noteworthy features. Notice $U(P)$ is of height 0, being the cyclic group $\{ xgx^{-1},1 \}$ of order 2. The orbits of the action of $U(P)$ on P are all singleton sets, except the orbit of z, which consists of the two maximal vertices $\{ z \}$ and $\{ xgx^{-1}z \}$. All the stabilizers of the action are trivial, except that $U(P)=\{ xgx^{-1},1 \}$ stabilizes $V_x=\{ x,xg \}$. The $\simeq$-class of x contains the cyclic orbit $V_{x^{-1}}$ and the noncyclic orbit V_x, (cf. cautionary note following lemma 4.17).

According to theorem A of section 6, we should be able to construct a graph of groups whose fundamental group is $\mathbf{U}(P)$. This graph will vary with the choices we make for a generating set of P.

Method 1: Select $X=\{ x,z \}$ as a spanning set. Then $\mathbf{B}=\{ V_{x^{-1}},V_{z^{-1}} \}$ is the basepoint set, and the fundamental group system is $\mathbf{G}=\{ <g>,1 \}$. We use the $< >$ notation to denote the subgroup generated by a set. Notice $A_x = <xgx^{-1}>$ and $A_z=1$. Hence $\mathbf{U}(P) \approx <g> * <xgx^{-1}> * F(x,z)/\ll x^{-1} \bullet xgx^{-1} \bullet x \bullet g \gg$. Double brackets of course denote the normal subgroup generated by a set. We construct the following graph of groups for $\mathbf{U}(P)$. To interpret the illustration, notice that edges in the maximal tree are hatched, trivial edge groups are not indicated, and edges not in the maximal tree are labeled with the elements of the spanning set.

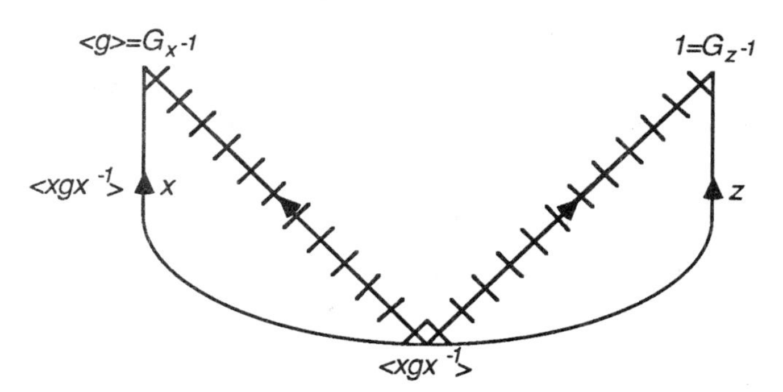

This graph is drawn according to the instructions given in the proof of theorem A in section 6. The basic ideas involved are

(i) to amalgamate appropriate fundamental groups of P with those of $U(P)$ along a maximal tree in the graph, and

(ii) to create an "HNN extension" for each element of the spanning set of P.

Method 2: Select $X=\{x^{-1},z\}$ as the spanning set. Then $\mathbf{B}=\{V_{x^{-1}},V_z\}$ and $\mathbf{G}=\{1_{x^{-1}},1_z\}$. (Recall from definition 4.20 that the fundamental groups of distinct simple vertices are distinct even if they coincide by abuse as subsets of P.) Since g is *maximal*, $A_{x^{-1}}=1$. The presentation of theorem 3 yields

$$\mathrm{U}(P)\approx<xgx^{-1}>*F(x^{-1},z)\approx\mathbb{Z}_2*\mathbb{Z}*\mathbb{Z}$$

The corresponding graph of groups given by theorem A is

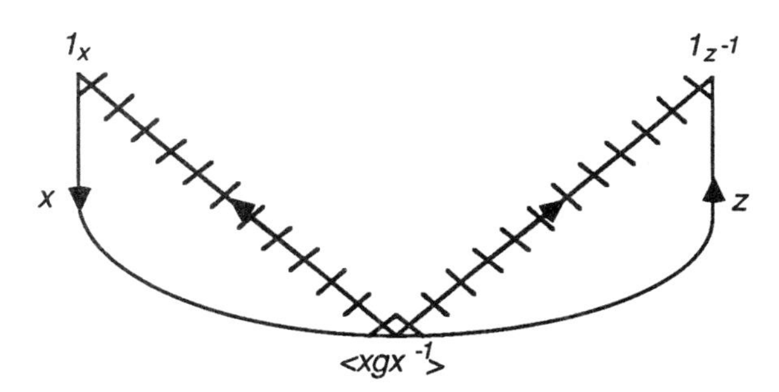

These two methods yield isomorphic groups. In fact the presentation of method 1 may be reduced to $\mathbb{Z}_2*\mathbb{Z}*\mathbb{Z}$ via a Tietze transformation eliminating the redundant generator g. Graphically, we see that the "unnecessary" edge and vertex groups have been removed.

5. Example: By definition, the *simple* pregroups have no non-trivial subgroups. By corollary 3.10, the only simple pregroup of height zero is the trivial group. Inductively, suppose that every simple pregroup of height $<n$ has free universal group.

Let P be simple of height n. By induction, the units of P have free universal group $\mathbf{U}(U)$. Since P has no cyclic maximal vertices, the fundamental group system for P is a set of distinguished trivial groups. Let $\mathbf{B}$ be a basepoint system for P, and let $X = \text{span}(\mathbf{B})$ be a spanning set based at $\mathbf{B}$. Using the notation of theorem 3, we see that $Q = FP(X) \oplus U$ and $R = \{ 1 \}$, so $\phi : \mathbf{U}(Q) \to \mathbf{U}(P)$ is an isomorphism. But $\mathbf{U}(Q) = F(X) * \mathbf{U}(U)$, a free product of free groups.

6. Corollary: The simple pregroups of finite height have free universal group. Furthermore, a free basis of $\mathbf{U}(P)$ can be found contained in P.

Part III
The relationship between pregroups and graphs of groups

6. A graph of groups structure for pregroups of finite height

In this section we prove theorem A, which is an algorithm for constructing a graph of groups from a given pregroup of finite height. As we have noted in section 5, the presentation of theorem 5.3 bears a strong resemblance to the presentation for a graph of groups. Think of $U(P)$ as a ball, and the fundamental group system **G** of P as a set of mushrooms growing out of $U(P)$. The spanning sets for the fundamental groups are then ropes tying the mushrooms down to their attaching groups inside of $U(P)$.

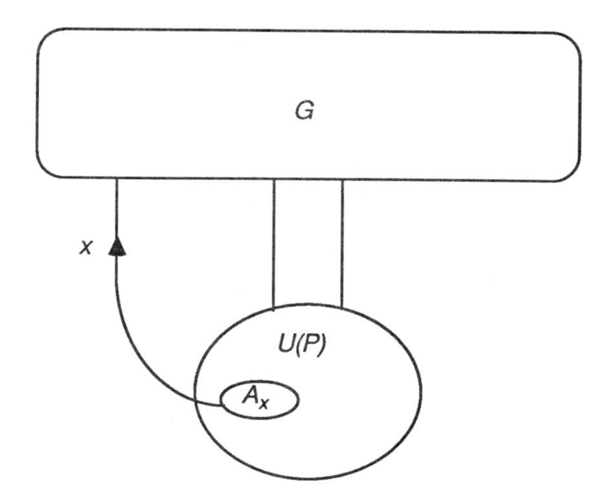

Repeat this same process for $U(P)$.

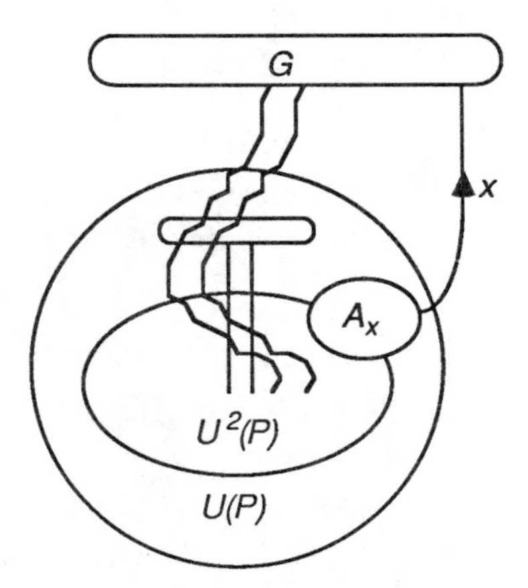

The problem is that the various fundamental groups and attaching groups of P, $U(P)$, $U^2(P) \cdots$ are not related to each other in an obvious way. For a long time I searched for a way to choose fundamental groups and attaching groups of a given level in such a way that there intersection with the next level down is contained in the fundamental groups chosen for that level. The following example shows this is not possible.

1. Example: Let P be the pregroup whose order tree is exhibited in figure 2, and whose partial multiplication table is indicated in figure 3. The group $\mathbf{U}(P)$ is generated by the subset $\{ a,b,\alpha,\beta,x,\gamma \}$ of P. The cyclic elements of P form the set

$$\{ a,\alpha,\alpha a,\beta b,xax^{-1},b,\beta,x^{-1}bx,1 \}$$

All the nontrivial cyclic elements are of order 2. P has height 3 and depth 2. (Pathology seems to occur when depth is strictly less than height.) P has two cyclic maximal vertices, with corresponding fundamental groups $G_\alpha = <\alpha,a>$ and $G_\beta = <\beta,b>$, both isomorphic to $\mathbb{Z}_2 \oplus \mathbb{Z}_2$.

The pregroup $U(P)$ is of height 1 and has two maximal vertices. In fact, $U(P)$ is a *partial group* (*cf*. Stallings [1966].) Both the maximal vertices of $U(P)$ are cyclic, with fundamental groups $G_a = <a,x^{-1}bx>$ and $G_b = <b,xax^{-1}>$ isomorphic to $\mathbb{Z}_2 \oplus \mathbb{Z}_2$. The pregroup $U^2(P)$ is just the trivial group 1. We make no attempt to prove that P is a pregroup, but Jim Shearer and I spent a **very long** evening with the computer and verified the pregroup axioms.

In figure 3 we have not indicated the value of any given product, as it may easily be deduced from context. A pair $(x,y) \in P \times P$ is given a ✓-mark $\Leftrightarrow$ $(x,y)_D$.

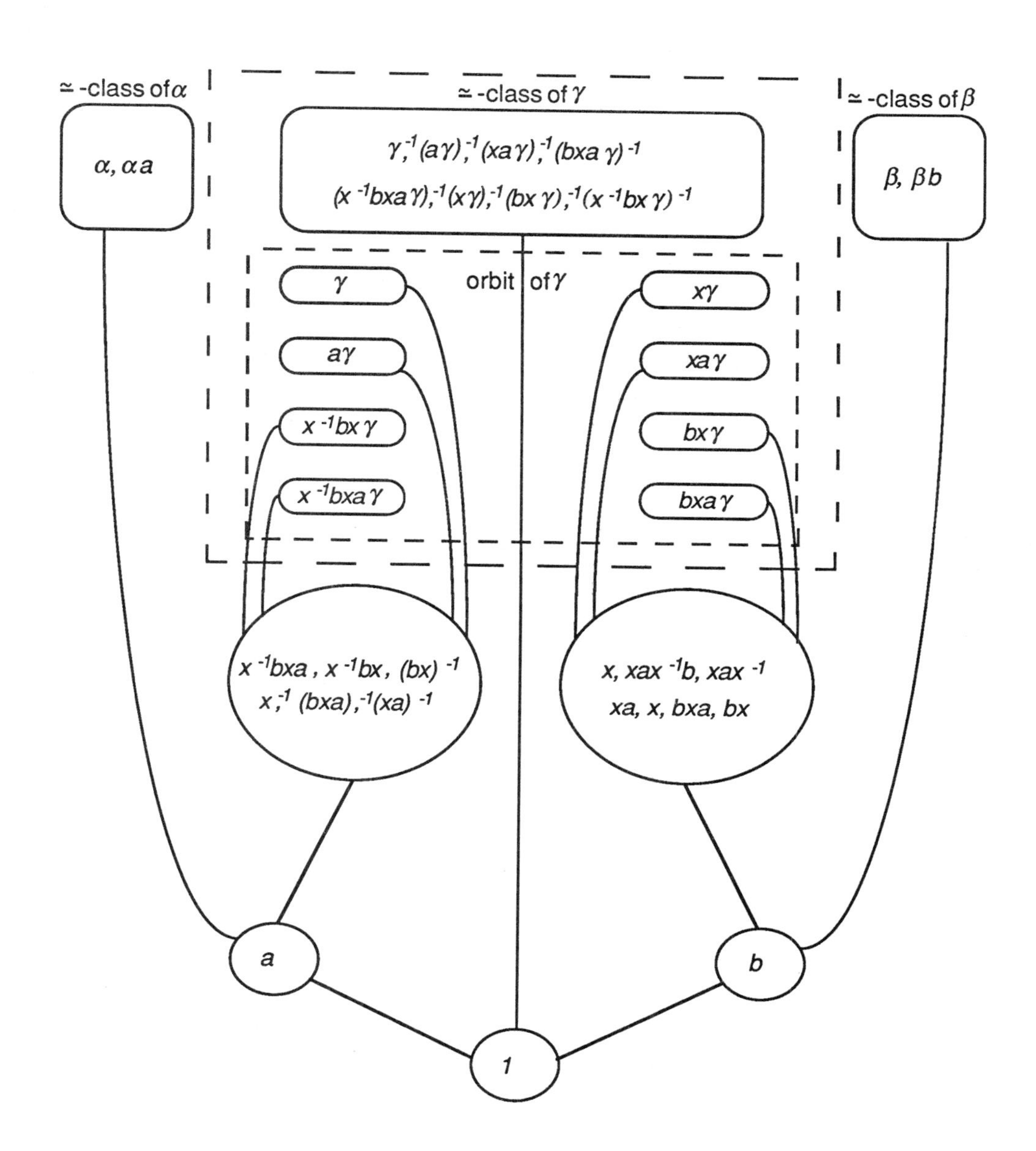

2. Figure: The order tree for P

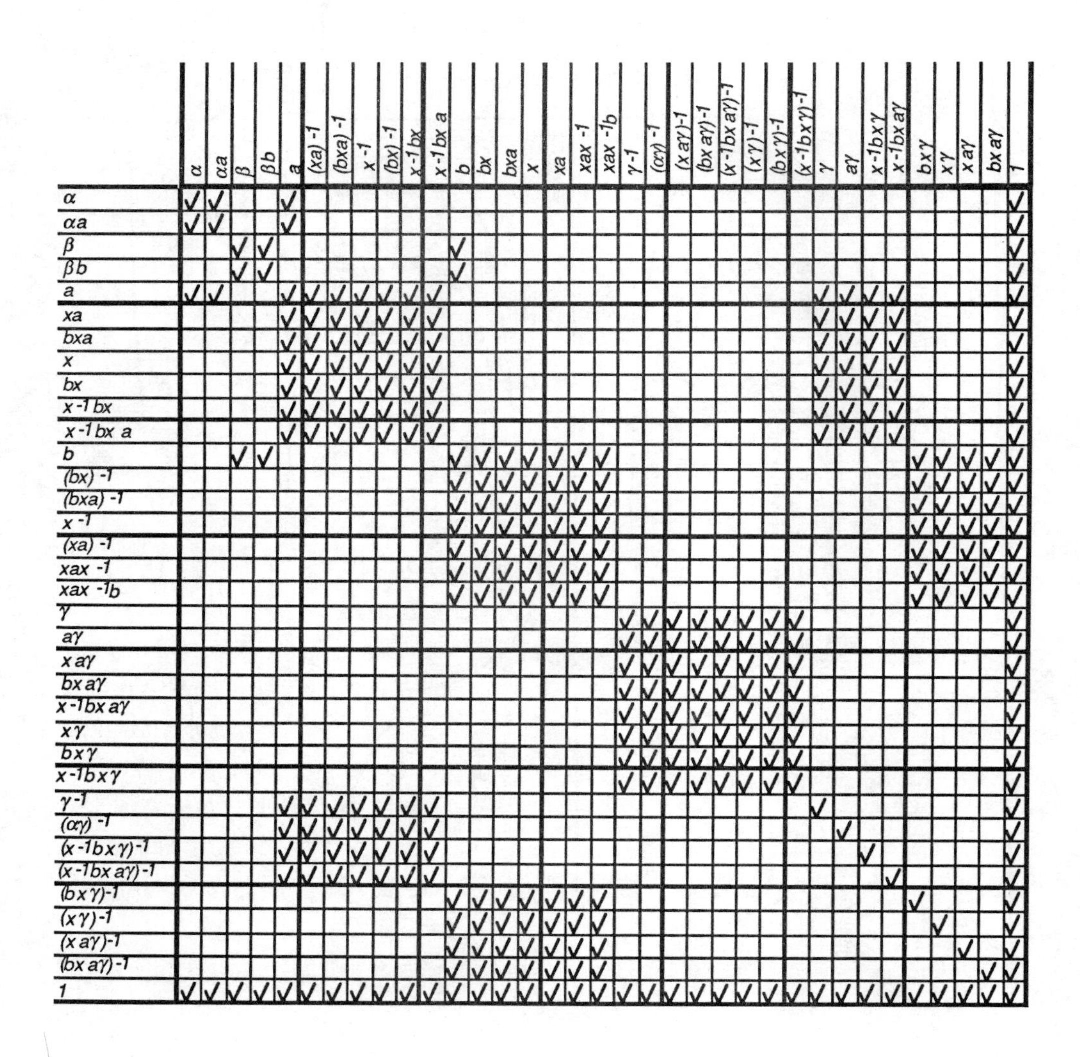

	α	αa	β	βb	a	$(xa)^{-1}$	$(bxa)^{-1}$	x^{-1}	$(bx)^{-1}$	$x^{-1}bx$	$x^{-1}bx\,a$	b	bx	bxa	x	xa	xax^{-1}	$xax^{-1}b$	γ^{-1}	$(\alpha\gamma)^{-1}$	$(xa\gamma)^{-1}$	$(bxa\gamma)^{-1}$	$(x^{-1}bxa\gamma)^{-1}$	$(x\gamma)^{-1}$	$(bx\gamma)^{-1}$	$(x^{-1}bx\gamma)^{-1}$	γ	$a\gamma$	$x^{-1}bx\gamma$	$x^{-1}bxa\gamma$	$bx\gamma$	$x\gamma$	$xa\gamma$	$bxa\gamma$	1
α	✓	✓			✓																														✓
αa	✓	✓			✓																														✓
β			✓	✓								✓																							✓
βb			✓	✓								✓																							✓
a	✓	✓			✓	✓	✓	✓	✓	✓	✓																✓	✓	✓	✓					✓
xa					✓	✓	✓	✓	✓	✓	✓																✓	✓	✓	✓					✓
bxa					✓	✓	✓	✓	✓	✓	✓																✓	✓	✓	✓					✓
x					✓	✓	✓	✓	✓	✓	✓																✓	✓	✓	✓					✓
bx					✓	✓	✓	✓	✓	✓	✓																✓	✓	✓	✓					✓
$x^{-1}bx$					✓	✓	✓	✓	✓	✓	✓																✓	✓	✓	✓					✓
$x^{-1}bx\,a$					✓	✓	✓	✓	✓	✓	✓																✓	✓	✓	✓					✓
b			✓	✓								✓	✓	✓	✓	✓	✓	✓													✓	✓	✓	✓	✓
$(bx)^{-1}$												✓	✓	✓	✓	✓	✓	✓													✓	✓	✓	✓	✓
$(bxa)^{-1}$												✓	✓	✓	✓	✓	✓	✓													✓	✓	✓	✓	✓
x^{-1}												✓	✓	✓	✓	✓	✓	✓													✓	✓	✓	✓	✓
$(xa)^{-1}$												✓	✓	✓	✓	✓	✓	✓													✓	✓	✓	✓	✓
xax^{-1}												✓	✓	✓	✓	✓	✓	✓													✓	✓	✓	✓	✓
$xax^{-1}b$												✓	✓	✓	✓	✓	✓	✓													✓	✓	✓	✓	✓
γ																			✓	✓	✓	✓	✓	✓	✓	✓									✓
$a\gamma$																			✓	✓	✓	✓	✓	✓	✓	✓									✓
$xa\gamma$																			✓	✓	✓	✓	✓	✓	✓	✓									✓
$bxa\gamma$																			✓	✓	✓	✓	✓	✓	✓	✓									✓
$x^{-1}bxa\gamma$																			✓	✓	✓	✓	✓	✓	✓	✓									✓
$x\gamma$																			✓	✓	✓	✓	✓	✓	✓	✓									✓
$bx\gamma$																			✓	✓	✓	✓	✓	✓	✓	✓									✓
$x^{-1}bx\gamma$																			✓	✓	✓	✓	✓	✓	✓	✓									✓
γ^{-1}					✓	✓	✓	✓	✓	✓	✓																✓								✓
$(\alpha\gamma)^{-1}$					✓	✓	✓	✓	✓	✓	✓																	✓							✓
$(x^{-1}bx\gamma)^{-1}$					✓	✓	✓	✓	✓	✓	✓																		✓						✓
$(x^{-1}bxa\gamma)^{-1}$					✓	✓	✓	✓	✓	✓	✓																			✓					✓
$(bx\gamma)^{-1}$												✓	✓	✓	✓	✓	✓	✓													✓				✓
$(x\gamma)^{-1}$												✓	✓	✓	✓	✓	✓	✓														✓			✓
$(xa\gamma)^{-1}$												✓	✓	✓	✓	✓	✓	✓															✓		✓
$(bxa\gamma)^{-1}$												✓	✓	✓	✓	✓	✓	✓																✓	✓
1	✓	✓	✓	✓	✓	✓	✓	✓	✓	✓	✓	✓	✓	✓	✓	✓	✓	✓	✓	✓	✓	✓	✓	✓	✓	✓	✓	✓	✓	✓	✓	✓	✓	✓	✓

3. Figure: The partial multiplication table for *P*

The fundamental groups of P and $U(P)$ are related to each other as follows. We use the mushroom notation described above.

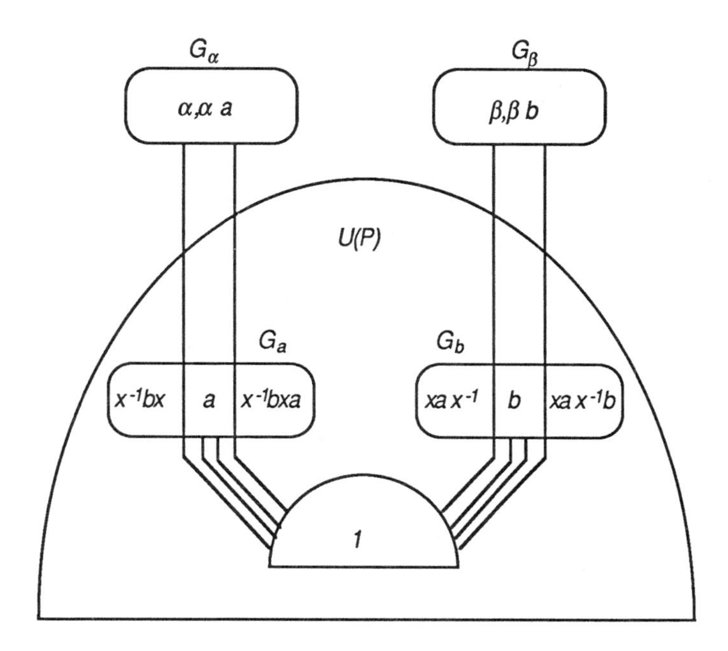

Consulting figure 2, we see that V_α and V_β are the unique P-maximal vertices in distinct $\approx$-classes of P. Hence any fundamental group system for P must contain both G_α and G_β. However, in $U(P)$, $a \approx x^{-1}$ and $b \approx x$, so that the fundamental group system for $U(P)$ is either $\{ G_a \}$ or $\{ G_b \}$. Suppose we select $\{ G_a \}$. The problem is that G_β is supposed to be amalgamated with G_b, but G_b was not picked for the generating set of $U(P)$. In fact there is no way to pick fundamental group systems $\mathbf{G}_P$ and $\mathbf{G}_{U(P)}$ for P and $U(P)$ respectively so that the groups in $\mathbf{G}_P$ "line up" over those in $\mathbf{G}_{U(P)}$. □

Nevertheless, if we conjugate G_b by x, then the fundamental groups line up in the larger group $\mathbf{U}(P)$. In fact, this phenomenon is perfectly general, and gives rise to a realization of $\mathbf{U}(P)$ as π_1(graph of groups). Modulo this difficulty, our proof of the relationship between pregroups and graphs of groups is to a large extent a matter of translating between two different notations, one for graphs of groups and one for generating sets of pregroups. First we fix the notation we shall use for graphs of groups. Our terminology follows Serre [1980] with occasional modifications which prevent confusion with the notation of this paper.

4. Definition: A *graph* Y consists of

(i) a set of *bases,* denoted B. These bases are the vertices of the graph.

(ii) a set of edges, denoted E,

(iii) initial and terminal base maps $E \rightarrow B \times B$ denoted $e \rightarrow (\iota(e), \tau(e))$,

(iv) a fixed point free involution of the edges of Y, denoted $e \rightarrow \bar{e}$, satisfying $\iota(e) = \tau(\bar{e})$.

If $e \in E$, then $\iota(e)$ is called the *initial base* of e, and $\tau(e)$ is called the *terminal base.*

5. Definition: Let Y be a connected graph. Suppose

(i) $\mathbf{G} = \{ G_v \}_{v \in B}$ is a set of groups indexed by the bases,

(ii) $\{ E_e \}_{e \in E}$ is a set of edge groups indexed by the edges, such that $E_e = E_{\bar{e}}$ for all $e \in E$. These edge groups will correspond exactly to the attaching groups of the spanning elements of the generating set.

(iii) $\{ \phi_e : E_e \rightarrow G_{\tau(e)} \}$ is a set of group monomorphisms, called the edge maps. The notation a^e means $a \in E_e$ and $\phi_e(a) = a^e \in G_{\tau(e)}$.

Then Y together with (i), (ii) and (iii) is a *graph of groups,* denoted $(\mathbf{G}, Y)$. □

Suppose $E' \subset E$ satisfies the condition $e \in E' \Rightarrow \bar{e} \in E'$. An *orientation* of E' is a subset $X \subset E'$ such that $e \in X \Rightarrow \bar{e} \notin X$ and for all $e \in E'$, either $e \in X$ or $\bar{e} \in X$. An orientation X of the edges E of Y is also called an *orientation of* Y. If X is an orientation of Y, and $x \in X$, we say that x is an *oriented edge.* If we already have a specific edge in mind, we denote the maps $\phi_{\bar{x}}$ and ϕ_x by $\alpha : E_x \rightarrow G_{\iota(x)}$ and $\beta : E_x \rightarrow G_{\tau(x)}$.

6. Definition: Let $(\mathbf{G}, Y)$ be a graph of groups, and let T be a maximal tree of Y. Let E be the edges of Y and let $E(T)$ be the edges of T. Let X be an orientation of $E - E(T)$. The *fundamental group of* $(\mathbf{G}, Y)$ *with respect to* T is $\pi_1(\mathbf{G}, Y, T)$:

$$\pi_1(\mathbf{G}, Y, T) = F(X) \underset{G \in \mathbf{G}}{*} G / \ll R \gg$$

where R is the set of relations

$$R = \{ \alpha(a) = \beta(a) \mid t \in E(T), a \in E_t \} \quad \cup \{ x^{-1} \alpha(a) x = \beta(a) \mid x \in Y, a \in A_x \} \quad \square$$

In example 1 we showed that the fundamental group systems cannot be *chosen* to line up so that all the upper mushrooms grow out of ones that are lower down. However, they can be *forced* to line up if we choose the upper mushroom carefully and then conjugate it by appropriate element of the spanning set of lower mushroom. To be sure, this process may force parts of the upper mushroom out of the pregroup P and into $\mathbf{U}(P) - P$, but as long as we keep track of the alterations we make as we proceed from each $U^i(P)$ to $U^{i-1}(P)$, we shall be able to construct an isomorphism between $\mathbf{U}(P)$ and the "mushrooms and ropes" graph of groups.

Recall the notion of the depth of a pregroup, defined directly above lemma 3.9.

7. Definition: Let P be a pregroup of depth $d \geq 1$. Let V be a maximal vertex of P. Let G_V be the fundamental group of V. Let A_V be the attaching group of V. For some m, $1 \leq m \leq d$, suppose that W is a cyclic maximal vertex of $U^m(P)$. Let G_W be the fundamental group of W with respect to $U^m(P)$. Then W *supports* V if $A_v \cap W \neq \emptyset$ and $A_V \subset G_W$. The number m is called the *support depth* of V. □

8. Lemma: Let P be a pregroup of depth $d \geq 1$. Let V be a maximal vertex of P. Then there exists a unique support depth m, $1 \leq m \leq d$, for V and a unique $U^m(P)$-maximal vertex W which supports V.

Proof: Let $m = \min(\text{depth of } a \mid s \in A_V)$. It follows that $A_V \subset U^m(P)$. Let W be a $U^m(P)$-maximal vertex containing an element of A_V. By theorem 4.10.iv, $A_V \subset G_W$.

9. Definition: Let $(\mathbf{G},X)$ be a generating set for P. Let V be a maximal vertex of P. Then $(\mathbf{G},X)$ *visits* V if $G_V \in \mathbf{G}$ or if there exists $x \in X$ such that $x \in V$.

10. Definition: Let P be a pregroup of depth d. A P-sequence is a sequence $(o_d, \ldots, o_i, \ldots, o_0)$ of objects satisfying the following property: each object $o_i = (\mathbf{G}_i, X_i)$ is a generating system for $U^i(P)$. □

Reading from left to right seems to correspond most naturally with traveling upwards; hence the reverse numbering scheme.

11. Definition: Given a pregroup P of depth d, define a *good* P-sequence inductively as follows.

(i) If $d=0$, the unique P-sequence $((P,\emptyset))$ is a good P sequence.

(ii) Let $(o_d, \ldots, o_0)$ be a P-sequence, such that $(o_d, \ldots, o_1)$ is a good $U(P)$ sequence. If for every P-maximal vertex V visited by o_0 there exists some i, $d \geq i \geq 1$, such that V is supported by some $U^i(P)$-maximal vertex visited by o_i, then $(o_d, \ldots, o_0)$ is a good P-sequence.

12. Lemma: Let P be a pregroup of depth d. Then P has a good P-sequence.

Proof: By induction, assume $d \geq 1$ and there is a good $U(P)$-sequence $(o_d, \ldots, o_1)$. Let $o_i = (\mathbf{G}_i, X_i)$ for $i = d, \ldots, 1$. Let $\tilde{\mathbf{B}}$ be a basepoint set for P, (definition 4.18). Given a P-maximal vertex $\tilde{V} \in \tilde{\mathbf{B}}$, let $\tilde{G}$ be the fundamental group of $\tilde{V}$. Our strategy will be to exchange the basepoint set $\tilde{\mathbf{B}}$ for a basepoint set $\mathbf{B}$ which "lines up" over $(o_d, \ldots, o_1)$.

Fix $\tilde{V} \in \tilde{\mathbf{B}}$. We shall exchange $\tilde{V}$ for a P-maximal vertex V with fundamental group G, defined as follows. Let m be the support depth of $\tilde{V}$, and let $\tilde{W}$ be the cyclic $U^m(P)$-maximal vertex supporting $\tilde{V}$.

If $m=d$, then $\tilde{V}$ is supported by the unique $U^d(P)$-maximal vertex. Set $V=\tilde{V}$ and $G=\tilde{G}$.

Now suppose $m<d$. By definition, each orbit O of the action of $U^{m+1}(P)$ on the $U^m(P)$-maximal vertices contains a vertex $V(O)$ visited by o_m. Let O be the $U^m(P)$ orbit containing $\tilde{W}$. Set $W=V(O)$. Hence there exists $u \in U^{m+1}(P)$ such that $u \in \tilde{W}=W$. Let $\tilde{g} \in \tilde{W} \cap G_{\tilde{V}}$. Then $u <_m \tilde{g} < \tilde{V}$, so by lemma 3.6, $u < \tilde{V}$. Let $V=u^{-1}\tilde{V}$. By lemma 4.14b, $G_V=u^{-1}G_{\tilde{V}}u$ and $G_W=u^{-1}G_{\tilde{W}}u$. Thus

$$\begin{aligned} A_V &= G_V \cap U(P) \\ &= u^{-1}(G_{\tilde{V}} \cap U(P))u \\ &= u^{-1}A_{\tilde{V}}u \\ &\subset u^{-1}G_{\tilde{V}}u \\ &= G_W. \end{aligned}$$

We conclude that for each $\tilde{V} \in \tilde{\mathbf{B}}$, there is a P-maximal vertex V such that the fundamental groups of these two vertices are in the same orbit. It follows that $\mathbf{G}_0 = \{ G_V \}_{V \in \mathbf{B}}$ is a fundamental group system for P. Moreover, V is supported by a vertex visited by o_m, for some $0 \leq m < d$.

We must now line up the spanning sets. Fix $V \in \mathbf{B}$ and let $\tilde{X}$ be an arbitrary spanning set based at V. Let $\tilde{x} \in \tilde{X}$ be such that $V_{\tilde{x}^{-1}}=V$. Arguing as above, let $u \in U(P)$ be such that $V_{u\tilde{x}}$ is supported by some W visited by o_m, for some $1 \leq m \leq d$. Notice $\tilde{x}^{-1}u^{-1} \approx \tilde{x}^{-1} \in V_{\tilde{x}^{-1}}=V$, and hence we may replace $\tilde{x}$ by $u\tilde{x}$. Do so for each element of the original spanning set for V, and for each $V \in \mathbf{B}$, forming a new spanning set X_0 based at $\mathbf{B}$. Setting $o_0=(\mathbf{G}_0, X_0)$, we see that $(o_d, \ldots, o_0)$ is a good P-sequence.

13. Definition: (The dot and vector notation). Suppose Q is a pregroup of depth d. Fix a good Q-sequence $(o_d, \ldots, o_0)$. Let each $o_i=(\mathbf{G}_i, X_i)$for $i=d, \ldots, 0$. Let $\mathbf{B}_i$ be the basepoint set for $U^i(Q)$ associated with o_i for each $i=d, \ldots, 0$. For each $v \in \mathbf{B}_i$, let span(v) be the spanning set for v such that span(v) $\subset X_i$. Fix some $i \geq 1$, let $G_v \in \mathbf{G}_i$, and let W support $v \in \mathbf{B}_i$. Let m be the support depth of v, so that W is a $U^{i+m}(Q)$-maximal vertex. Since $(o_d, \ldots, o_0)$ is a good spanning set, o_{i+m} visits W by property (ii) of definition 11. Therefore there exists $\dot{v} \in \mathbf{B}_{i+m}$ and $\vec{v} \in X_{i+m} \cup \{ 1 \}$ such that

$$\begin{aligned} &\vec{v} \in W \cap \text{span}(\dot{v}) && \text{if } W \notin \mathbf{G}_{i+m}, \text{ and} \\ &\vec{v} = 1 && \text{if } W \in \mathbf{G}_{i+m}. \end{aligned}$$

This is the *dot and vector* notation for a *basepoint* $v \in \mathbf{B}_i$ with respect to

$(o_d, \ldots, o_0)$.

Similarly, given $x \in X_i$, let W support V_x, with support depth m. Define $\dot{x} \in \mathbf{B}_{i+m}$ and $\vec{x} \in X_{i+m} \cup \{ 1 \}$ such that

$$\vec{x} \in W \cap \operatorname{span}(\dot{x}) \qquad \text{if } W \notin \mathbf{G}_{i+m},$$
$$\vec{x} = 1 \qquad \text{if } W \in \mathbf{G}_{m+i}.$$

This is the *dot and vector* notation for a *spanning element* $x \in X_i$, with respect to $(o_d, \ldots, o_0)$. □

As an exercise, one may construct a good spanning set for the pregroup of example 1, and determine which elements of the pregroup are the dot and the vector of a given basepoint or spanning element.

The following two lemmas allow us to "line up" a given fundamental group over one at a lower level, and also to conjugate an attaching subgroup into a fundamental group at a lower level. The precise application of these lemmas will be revealed at a crucial point in the proof of theorem 17 below.

14. Lemma: Let P be of finite depth d. Let $(o_d, \ldots, o_0)$ be a good P-sequence. Let $v \in \mathbf{B}_0$, the basepoint set of o_0. Then $\vec{v}^{-1} A_v \vec{v} \subset G_{\dot{v}}$.

Proof: Unravel the notation and use lemma 4.14.a. □

15. Lemma: Let P be of finite depth d. Let $(o_d, \ldots, o_0)$ be a good P-sequence. Let $x \in X_0$, the spanning set for o_0. Then $\vec{x}^{-1} A_x \vec{x} \subset G_{\dot{x}}$

Proof: Same as lemma 14. □

16. Definition: (The inductive hypothesis) Let $d \geq j \geq 0$. Let Q be a pregroup of depth d. Let $U = U^j(Q)$. Fix a good Q-sequence $(o_d, \ldots, o_0)$, $o_i = (\mathbf{G}_i, X_i)$, with corresponding basepoint set $\mathbf{B}_i$, $i = d, \ldots, 0$. The *Inductive hypothesis for Q at level j with respect to* $(o_d, \ldots, o_0)$ is as follows.

(i) There exists a graph Y with edges E and maximal tree T with edges $E(T)$. The union of the spanning sets $X = \bigcup_{i=j}^{d} X_i$ is a subset of E. Moreover, X is an orientation for $E - E(T)$. The set of bases is $\mathbf{B} = \bigcup_{i=j}^{d} \mathbf{B}_i$. The maximal tree T has an orientation $T^+ \subset E(T)$ enjoying the following properties:

 (a) there exists a bijection $\mathbf{v}: T^+ \rightarrow \{ v \in \mathbf{B} \mid v \notin \mathbf{B}_d \}$,

 (b) for all $v \in \mathbf{B}$ such that $v \notin \mathbf{B}_d$, we have $\tau(\mathbf{v}^{-1}(v)) = v$.

(ii) There exist

 (a) for each $v \in \mathbf{B}$, a group H_v and a group isomorphism $\tilde{v}: G_v \rightarrow H_v$, and there exists

(b) a graph of groups $(\mathbf{H},Y)$ with base group H_v,

(c) a set map $\mathbf{u}:\mathbf{B}\rightarrow \mathrm{U}(P)$, and

(d) a group isomorphism $\Phi:\pi_1(\mathbf{H},Y,T)\rightarrow \mathrm{U}(P)$, enjoying the property

$$\text{for all } v\in \mathbf{B}, \text{ for all } g\in G_v,\ \Phi\circ \tilde{v}(g)=\mathbf{u}(v)^{-1}g\mathbf{u}(v)$$

17. Theorem: Let Q be a pregroup of height $d\geq 1$. Let $d>j\geq 0$. Fix a good Q-sequence $(o_d,\ldots,o_0)$, $o_i=(\mathbf{G}_i,X_i)$, with corresponding basepoint set $\mathbf{B}_i$, $i=d,\ldots,0$. Suppose Q satisfies the inductive hypothesis at level j with respect to $(o_d,\ldots,o_0)$. Then Q satisfies the inductive hypothesis at level $j-1$ with respect to $(o_d,\ldots,o_0)$.

Proof: Let P be the pregroup $U^{j-1}(Q)$, and let $U=U^j(Q)$. As we are not concerned with $o_{j-2},\ldots,o_0$ we assume that $j=1$ and hence $P=Q$. Thus $(\mathbf{G}_0,X_0)$ is a generating system for P, and P satisfies the inductive hypothesis at level 1. Therefore we assume the existence of

$$Y,\ E,\ T,\ X,\ \mathbf{B},\ \mathbf{v},\ \{\ H_v,\ \tilde{v}\ |\ v\in\mathbf{B}\ \},\ (\mathbf{H},Y),\ \mathbf{u}, \text{ and } \Phi$$

satisfying the inductive hypotheses 16.i and 16.ii for P at level 1.

We extend $(\mathbf{H},Y)$ to a graph of groups $(\hat{\mathbf{H}},\hat{Y})$ as follows. Let $\hat{X}=X\cup X_0$ and let $\hat{\mathbf{B}}=\mathbf{B}\cup\mathbf{B}_0$. Let T_0 be a set that is disjoint from $E\cup X_0$ and in 1-1 correspondence with $\mathbf{B}_0$. Set $\hat{T}^+=T^+\cup T_0$. Extend $\mathbf{v}:T^+\rightarrow\{\ v\in\mathbf{B}\ |\ v\notin\mathbf{B}_d\ \}$ to a bijection $\hat{\mathbf{v}}:\hat{T}^+\rightarrow\{\ v\in\hat{\mathbf{B}}\ |\ v\notin\mathbf{B}_d\ \}$. Let X_0^- and T_0^- be oppositely oriented edges, so that $\hat{E}=E^+\cup X_0\cup X_0^{-1}\cup T_0\cup T_0^{-1}$ is the set of edges of $\hat{Y}$. Let $\hat{T}$ be the maximal tree of $\hat{Y}$ with orientation $\hat{T}^+$.

Fix an edge $t\in T_0$. We define the initial and terminal vertex maps of t as follows. Let $v=\mathbf{v}(t)$. Then $\iota(t)=\dot{v}$, and $\tau(t)=v$. This definition is consistent with inductive hypothesis 16.i.b.

Consider an edge $x\in X_0$. Let $v_{x^{-1}}\in\mathbf{B}_0$ be such that $x\in\mathrm{span}(v_{x^{-1}})$. We define the initial and terminal vertex maps as follows: $\iota(x)=\dot{v}_{x^{-1}}$, and $\tau(x)=v_{x^{-1}}$.

For each $v\in\mathbf{B}_0$, let H_v be a group isomorphic to G_v via $\tilde{v}:G_v\rightarrow H_v$. Set $\hat{\mathbf{H}}=\mathbf{H}\cup\{\ H_v\ |\ v\in\mathbf{B}_0\ \}$. To give $(\hat{\mathbf{H}},\hat{Y})$ the structure of a graph of groups we must specify the edge groups of the newly created edges $e\in X_0\cup\mathbf{B}_0$ and the group monomorphisms $\alpha:E_e\rightarrow H_{\iota(e)}$ and $\beta:E_e\rightarrow H_{\tau(e)}$. We set

$$E_t=A_t=A_v=G_v\cap U \qquad \text{for } t\in T_0,\ v=\hat{\mathbf{v}}(t)$$
$$E_x=A_x=G_x\cap U \qquad \text{for } x\in X_0$$

In other words, the edge groups are exactly the attaching groups of section 4. Henceforth we shall stick to the attaching group notation. The families of group monomorphisms α and β are extended to $X_0 \cup T_0$ as follows.

	$a \in A_t,\ t \in T_0$	$a \in A_x,\ x \in X_0$
$\alpha(a)=$	$\tilde{\dot{v}}(\vec{v}^{-1} a \vec{v}),\ \hat{v} = \mathbf{v}(t)$	$\tilde{\dot{x}}(\vec{x}^{-1} a \vec{x})$
$\beta(a)=$	$\tilde{v}(a),\ v = \hat{\mathbf{v}}(t)$	$\tilde{v}(x^{-1} a x),\ v = v_{x^{-1}}$

Explanatory Remarks: (1) The symbol $\tilde{\dot{v}}$ denotes the group isomorphism $\tilde{\dot{v}}: G_{\dot{v}} \to H_{\dot{v}}$. Thus we first apply "dot" to v, then "tilde" to $\dot{v}$. Similar comments apply to $\tilde{\dot{x}}$.

(2) Let $t \in T_0$. Let $v = \hat{\mathbf{v}}(t)$. Then lemma 14 implies that $\vec{v}^{-1} A_t \vec{v} \subset G_{\dot{v}}$. It follows that $\alpha(a)$ is well defined for $a \in A_t$. Similarly, lemma 15 implies that $\alpha(a)$ is well defined for $a \in A_x$, $x \in X_0$.

(3) Let $t \in T$, $v = \hat{\mathbf{v}}(t)$ and $a \in A_t$. Then $A_t \subset G_v$ by definition, so $\beta(a)$ is well defined. Let $x \in X_0$, $a \in A_x$. By remark 5.2, $(x^{-1}, a, x) \in G_{x^{-1}}$. Hence $\beta(a)$ is well defined.

(4) We have shown that α and β are defined as compositions of group isomorphisms and group monomorphisms; hence α and β are monomorphisms. □

We have now given $(\hat{\mathbf{H}}, \hat{Y}, \hat{T})$ the structure of a graph of groups. From definition 6, $\pi_1(\hat{\mathbf{H}}, \hat{Y}, \hat{T})$ is the group

$$\pi_1(\mathbf{H}, Y, T) \underset{v \in \mathbf{B}_0}{*} H_v * F(X_0) / <<R>>$$

where R is the set of relations

$$\begin{array}{lll} R = & \{\ \alpha(a) = \beta(a)\ \mid & a \in A_v,\ v \in \mathbf{B}_0\ \} \\ \cup & \{\ x^{-1}\alpha(a)x = \beta(a)\ \mid & a \in A_x,\ x \in X_0\ \} \end{array}$$

and $<<R>>$ is the normal subgroup of $\pi_1(\hat{\mathbf{H}}, \hat{Y}, \hat{T})$ generated by R. To complete inductive step 16.ii we extend $\Phi: \pi_1(\mathbf{H}, Y, T) \to \mathbf{U}(U)$ to a group isomorphism $\hat{\Phi}: \pi_1(\hat{\mathbf{H}}, \hat{Y}, \hat{T}) \to \mathbf{U}(P)$. Our strategy is to first define a map, also called $\hat{\Phi}$, on $\pi_1(\mathbf{H}, Y, T) \underset{v \in \mathbf{B}_0}{*} H_v * F(X_0)$ and then show that $R \subset \ker \hat{\Phi}$. Hence the induced map $\hat{\Phi}: \pi_1(\hat{\mathbf{H}}, \hat{Y}, \hat{T}) \to \mathbf{U}(P)$ is well defined. To show $\hat{\Phi}$ is an isomorphism we construct the inverse map $\Psi: \mathbf{U}(P) \to \pi_1(\hat{\mathbf{H}}, \hat{Y}, \hat{T})$ by a similar method.

We must also verify that $\hat{\Phi}$ satisfies inductive hypothesis 16.ii.d. To this end, extend $\mathbf{u}: \mathbf{B} \to \mathbf{U}(P)$ to $\hat{\mathbf{u}}: \hat{\mathbf{B}} \to \mathbf{U}(P)$ by the formula $\hat{\mathbf{u}}(v) = \vec{v}\mathbf{u}(\dot{v})$, $v \in \mathbf{B}_0$. For convenience, we also define $\hat{\mathbf{u}}(x) = \vec{x}\mathbf{u}(\dot{x})$ for $x \in X_0$.

Define a group homomorphism $\hat{\Phi}: \pi_1(\mathbf{H}, Y, T) \underset{v \in \mathbf{B}_0}{*} H_v * F(X_0) \to \mathbf{U}(P)$ as follows.

$$\hat{\Phi}(u)=\Phi(u) \qquad \text{for } u\in\pi_1(\mathbf{H},Y,T)$$
$$\hat{\Phi}(h)=\hat{\mathbf{u}}(v)^{-1}\tilde{v}^{-1}(h)\hat{\mathbf{u}}(v) \qquad \text{for } h\in H_v$$
$$\hat{\Phi}(x)=\hat{\mathbf{u}}(x)^{-1}x\hat{\mathbf{u}}(v) \qquad \text{for } x\in X_0,\ v=v_{x^{-1}}$$

This definition of $\hat{\Phi}$ is in accordance with inductive hypothesis 16.ii.d. We claim that $\hat{\Phi}$ factors through $\pi_1(\hat{\mathbf{H}},\hat{Y},\hat{T})$. First we check the relation $\alpha(a)=\beta(a)$ for $a\in A_v$ and $v\in\mathbf{B}_0$. Notice $\alpha(a)\in H_{\dot{v}}\subset\pi_1(\mathbf{H},Y,T)$, so that

$$\begin{aligned}
\hat{\Phi}(\alpha(a)) &= \Phi(\alpha(a)) && \text{definition of } \hat{\Phi}\\
&= \Phi(\tilde{v}(\vec{v}^{-1}a\vec{v})) && \text{definition of } \alpha\\
&= \mathbf{u}(\dot{v})^{-1}\vec{v}^{-1}a\vec{v}\,\mathbf{u}(\dot{v}) && \text{inductive hypothesis 16.ii.d}\\
&= \hat{\mathbf{u}}(v)^{-1}a\hat{\mathbf{u}}(v) && \text{definition of } \hat{\mathbf{u}}(v)
\end{aligned}$$

Moreover, $\beta(a)\in H_v$, so that

$$\begin{aligned}
\hat{\Phi}(\beta(a)) &= \hat{\mathbf{u}}(v)^{-1}\tilde{v}^{-1}(\beta(a))\hat{\mathbf{u}}(v) && \text{definition of } \hat{\Phi}\\
&= \hat{\mathbf{u}}(v)^{-1}\tilde{v}^{-1}(\tilde{v}(a))\hat{\mathbf{u}}(v) && \text{definition of } \beta\\
&= \hat{\Phi}(\alpha(a)) && \text{from above}
\end{aligned}$$

Now suppose we have a relation of the form $x^{-1}\alpha(a)x=\beta(a)$ for $a\in A_x$ and $x\in X_0$. Let $b=xax^{-1}\in G_v$, $v=v_{x^{-1}}$. Then

$$\begin{aligned}
\Phi(x^{-1}\alpha(a)x) &= \hat{\Phi}(x^{-1}\tilde{x}(\vec{x}^{-1}a\vec{x})x)\\
&= \hat{\Phi}(x^{-1})\Phi(\tilde{x}(\vec{x}^{-1}a\vec{x}))\hat{\Phi}(x)\\
&= \hat{\mathbf{u}}(v)^{-1}x^{-1}\hat{\mathbf{u}}(x)\mathbf{u}(\dot{x})^{-1}(\vec{x}^{-1}a\vec{x})\mathbf{u}(\dot{x})\hat{\mathbf{u}}(x)^{-1}x\hat{\mathbf{u}}(v)\\
&= \hat{\mathbf{u}}(v)^{-1}x^{-1}\hat{\mathbf{u}}(x)\hat{\mathbf{u}}(x)^{-1}a\hat{\mathbf{u}}(x)\hat{\mathbf{u}}(x)^{-1}x\hat{\mathbf{u}}(v)\\
&= \hat{\mathbf{u}}(v)^{-1}x^{-1}ax\hat{\mathbf{u}}(v)\\
&= \hat{\mathbf{u}}(v)^{-1}b\hat{\mathbf{u}}(v)
\end{aligned}$$

On the other hand

$$\begin{aligned}
\hat{\Phi}(\beta(a)) &= \hat{\Phi}(\tilde{v}(x^{-1}ax))\\
&= \hat{\mathbf{u}}(v)^{-1}x^{-1}ax\hat{\mathbf{u}}(v)\\
&= \hat{\mathbf{u}}(v)^{-1}b\hat{\mathbf{u}}(v)
\end{aligned}$$

We conclude that $\hat{\Phi}$ factors through $\pi_1(\hat{\mathbf{H}},\hat{Y},\hat{T})$, and henceforth $\hat{\Phi}$ will denote the induced homomorphism $\hat{\Phi}:\pi_1(\hat{\mathbf{H}},\hat{Y},\hat{T})\to \mathrm{U}(P)$. It is easy to see that $U(P)\cup X_0\ \bigcup_{v\in\mathbf{B}_0} G_v\subset \mathrm{im}\ \hat{\Phi}$, so theorem 5.3 implies that $\hat{\Phi}$ is actually an epimorphism.

To construct an inverse for $\hat{\Phi}$, consider the following diagram.

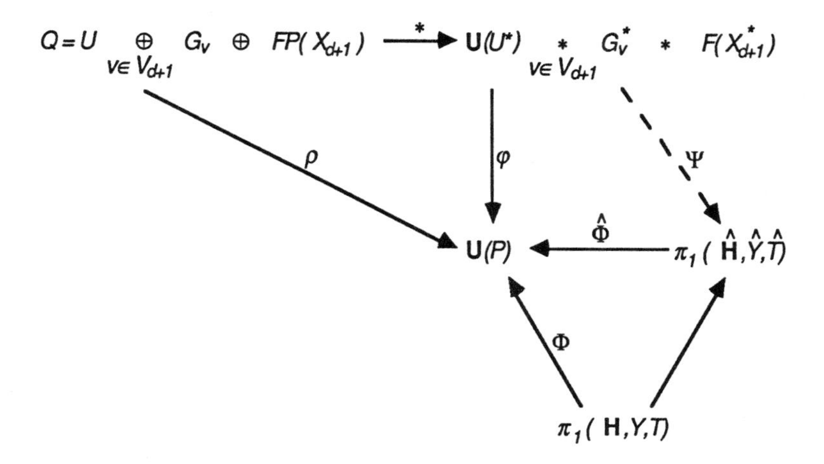

In order to be precise, we replace the pregroup $U \underset{v \in \mathbf{B}_0}{\oplus} G_v \oplus FP(X_d)$ with a starred copy. The map ρ restricted to any component of Q is a pregroup injection. Set $X_0^* = * \circ \rho^{-1}(X_0)$, $G_v^* = * \circ \rho^{-1}(G_v)$, and $A_v^* = * \circ \rho^{-1}(A_v)$, for $v \in \mathbf{B}_0$. For each $x \in X_0$, let $x^* = * \circ \rho^{-1}(x)$, and similarly let $g^* = * \circ \rho^{-1}(g)$ for each $g \in G_v$, $v \in \mathbf{B}_0$.

The map ϕ is the projection of theorem 5.3. For any $q \in Q$ we have the identity $\rho(q) = \phi(q^*)$. By theorem 5.3, the kernel of ϕ is normally generated by the relations

$$\{\ a^* = b^* \ \mid\ a^* \in U^*,\ b^* \in G_v^*,\ v \in \mathbf{B}_0,\ \phi(a^*) = \phi(b^*)\ \}$$

$$\cup \{\ x^{*-1} a^* x^* = b^* \ \mid\ x^* \in X_0^*,\ a^* \in A_x^*,\ \phi(x^{*-1} a^* x^*) = \phi(b^*)^{-1} \in G_v,\ v = v_{x^{-1}}\ \}$$

Our strategy is to construct a group homomorphism Ψ as indicated and show that $\ker \Phi \subset \ker \Psi$. We then show that the induced map $\Psi : \mathbf{U}(P) \to \pi_1(\hat{\mathbf{H}}, \hat{Y}, \hat{T})$ satisfies $\Psi\Phi = 1_{\pi_1(\hat{\mathbf{H}}, \hat{Y}, \hat{T})}$. Since Φ is onto, it follows that Φ is an isomorphism, completing the proof of inductive step (iii).

Set

$$\begin{aligned}
&\Psi(u^*) = \Phi^{-1}\phi(u^*) && \text{if } u^* \in U^* \\
&\Psi(g^*) = \Phi^{-1}(\hat{\mathbf{u}}(v))\tilde{v}\,(\phi(g^*))\Phi^{-1}(\hat{\mathbf{u}}(v)^{-1}) && \text{if } g^* \in G_v^*,\ v \in \mathbf{B}_0 \\
&\Psi(x^*) = \Phi^{-1}(\hat{\mathbf{u}}(x))x\,\Phi^{-1}(\hat{\mathbf{u}}(v)^{-1}) && \text{if } x^* \in X_0^*,\ v = v_{x^{-1}}.
\end{aligned}$$

The computations to show that Ψ factors through $\mathbf{U}(P)$ are as follows. Consider a relation of the form $a^* = b^*$, such that $a^* \in U^*$, $b^* \in G_v^*$, $\phi(a^*) = \phi(b^*) = a \in U$. Clearly

$\Psi(a^*)=\Phi^{-1}(\phi a^*)=\Phi^{-1}(a)$. The point is to make sure certain elements are in $\pi_1(\mathbf{H},Y,T)$, considered as a subgroup of $\pi_1(\hat{\mathbf{H}},\hat{Y},\hat{T})$. Observe that $a\in G_v\cap U=A_v$. Thus $\alpha(a)$ is defined, and $\alpha(a)\in H_{\dot{v}}\subset\pi_1(\mathbf{H},Y,T)$. Thus $\Phi(\alpha(a))\in\mathbf{U}(P)$ is defined. Since $\alpha(a)=\beta(a)=\tilde{v}(a)$, it follows that $\Phi(\tilde{v}(a))$ is a well defined element of $\mathbf{U}(U)$. Thus

$$\begin{aligned}
\Psi(b^*) &=\Phi^{-1}(\hat{\mathbf{u}}(v))\tilde{v}\,(\phi(b^*))\Phi^{-1}(\hat{\mathbf{u}}(v)^{-1})\\
&=\Phi^{-1}(\hat{\mathbf{u}}(v))\tilde{v}\,(a)\Phi^{-1}(\hat{\mathbf{u}}(v)^{-1})\\
&=\Phi^{-1}(\hat{\mathbf{u}}(v)\Phi^{-1}(\Phi(\tilde{v}\,(a))\Phi^{-1}(\hat{\mathbf{u}}(v)^{-1})\\
&=\Phi^{-1}(\hat{\mathbf{u}}(v)\Phi(\alpha(a))\hat{\mathbf{u}}(v)^{-1})\\
&=\Phi^{-1}(\hat{\mathbf{u}}(v)[\Phi\circ\tilde{\dot{v}}(\vec{v}^{-1}a\vec{v})]\hat{\mathbf{u}}(v)^{-1})\\
&=\Phi^{-1}(\hat{\mathbf{u}}(v)\mathbf{u}(\dot{v})^{-1}(\vec{v}^{-1}a\vec{v})\mathbf{u}(\dot{v})\hat{\mathbf{u}}(v)^{-1})\\
&=\Phi^{-1}(a)\\
&=\Psi(a^*)
\end{aligned}$$

Now consider a relation of the form $x^{*-1}a^*x^*=b^*$. In $\mathbf{U}(P)$ this implies the relation $x^{-1}ax=b$.

$$\begin{aligned}
\Psi(x^{*-1}a^*x^*) &=\Phi^{-1}(\hat{\mathbf{u}}(v))x^{-1}\Phi^{-1}(\hat{\mathbf{u}}(x)^{-1})\Phi^{-1}(a)\Phi^{-1}(\hat{\mathbf{u}}(x))x\,\Phi^{-1}(\hat{\mathbf{u}}(v)^{-1})\\
&=\Phi^{-1}(\hat{\mathbf{u}}(v))x^{-1}\Phi^{-1}(\hat{\mathbf{u}}(x)^{-1}a\,\hat{\mathbf{u}}(x))x\,\Phi^{-1}(\hat{\mathbf{u}}(v)^{-1})
\end{aligned}$$

Notice that $\vec{x}^{-1}a\vec{x}\in G_{\dot{x}}$, so that by the inductive hypothesis

$$\begin{aligned}
\tilde{\dot{x}}(\vec{x}^{-1}a\vec{x}) &=\Phi^{-1}(\mathbf{u}(\dot{x})^{-1}(\vec{x}^{-1}a\vec{x})\mathbf{u}(\dot{x}))\\
&=\Phi^{-1}(\hat{\mathbf{u}}(x)^{-1}a\,\hat{\mathbf{u}}(x)).
\end{aligned}$$

Thus $\Psi(x^{*-1}a^*x^{*-1})=\Phi^{-1}(\hat{\mathbf{u}}(v))x^{-1}\dot{x}(\vec{x}^{-1}a\vec{x})x\,\Phi^{-1}(\hat{\mathbf{u}}(v)^{-1})$. But $\tilde{\dot{x}}(\vec{x}^{-1}a\vec{x})=\alpha(a)$, where $\alpha{:}A_x\rightarrow H_{\iota(x)}$. Since $x^{-1}\alpha(a)x=\beta(a)=\tilde{v}(x^{-1}ax)=\tilde{v}(b)$,

$$\begin{aligned}
\Psi(x^{*-1}a^*x^*) &=\Phi^{-1}(\hat{\mathbf{u}}(v))\tilde{v}\,(b)\Phi^{-1}(\hat{\mathbf{u}}(v)^{-1})\\
&=\Psi(b^*).
\end{aligned}$$

Thus ϕ induces a map $\Psi{:}\mathbf{U}(P)\rightarrow\pi_1(\hat{\mathbf{H}},\hat{Y},\hat{T})$. We show that $\Psi\hat{\Phi}=1_{\pi_1(\hat{\mathbf{H}},\hat{Y},\hat{T})}$, completing the proof. For $u\in\pi_1(\mathbf{H},Y,T)$, clearly $\Psi\hat{\Phi}(u)=\Phi^{-1}\Phi(u)=u$. If $h\in H_v$, and $v\in X_0$, then

$$\begin{aligned}\Psi\hat{\Phi}(h) &= \Psi(\hat{\mathbf{u}}(v)^{-1}\tilde{v}^{-1}(h)\hat{\mathbf{u}}(v)) \\ &= \Psi(\hat{\mathbf{u}}(v)^{-1})\Psi(\tilde{v}^{-1}(h))\Psi(\hat{\mathbf{u}}(v)) \\ &= \Phi^{-1}(\hat{\mathbf{u}}(v)^{-1})\Phi^{-1}(\hat{\mathbf{u}}(v)^{-1})\tilde{v}\,\tilde{v}^{-1}(h)\Phi^{-1}(\hat{\mathbf{u}}(v)^{-1})\Phi^{-1}(\hat{\mathbf{u}}(v)) \\ &= h.\end{aligned}$$

Finally, if $x \in X_0$, then

$$\begin{aligned}\Psi\hat{\Phi}(x) &= \Psi(\hat{\mathbf{u}}(x)^{-1})x\Psi(\hat{\mathbf{u}}(v)) \\ &= \Phi^{-1}(\hat{\mathbf{u}}(x)^{-1})\Phi^{-1}(\hat{\mathbf{u}}(x))x\Phi^{-1}(\hat{\mathbf{u}}(v)^{-1})\Phi^{-1}(\hat{\mathbf{u}}(v)) \\ &= x.\end{aligned}$$

We conclude that

$$\hat{Y},\ \hat{E},\ \hat{T},\ \hat{X},\ \hat{\mathbf{B}},\ \hat{\mathbf{v}},\ \{\ H_v,\ \tilde{v}\ \mid\ v \in \hat{\mathbf{B}}\ \},\ (\hat{\mathbf{H}},\hat{Y}),\ \hat{\mathbf{u}},\text{ and }\hat{\Phi}$$

satisfy the inductive hypotheses for P at level 0. □

We have determined the structure of $\mathbf{U}(P)$ for P of finite depth, or equivalently, for P of finite height. The notion of height derives from the tree order for P. On the other hand, the notion of depth exists by virtue of theorem 3.4. Accordingly, we state theorem A in terms of height, the more primitive notion.

Theorem A: Let P be a pregroup of finite height. Let d be the depth of P. Let $\mathbf{G}$ be the union of the fundamental group systems for each $U^i(P)$, $i = 0, \ldots, d$. Then P is a pregroup structure for a group of the form $\pi_1(\mathbf{G}, Y)$. The vertex groups of Y are in 1-1 correspondence with the elements of $\mathbf{G}$, and corresponding groups are isomorphic. The graph Y is such that the oriented edges of the complement of a maximal tree in Y are in one to one correspondence with the union of the spanning sets for the fundamental groups in $\mathbf{G}$.

Proof: If P is of depth 0, then P is a group and the corresponding graph of groups consists of a single vertex with P as vertex group. Hence every pregroup of depth 0 satisfies the inductive hypotheses of definition 16. If P is of depth $d \geq 1$, then it follows that P satisfies the inductive hypotheses a level d. Apply theorem 17 d times to create the desired graph of groups.

18. Corollary If P is a finite pregroup, then $\mathbf{U}(P)$ is the fundamental group of a finite graph of finite groups. □

At this point it is appropriate to quote a theorem due to Karrass, Pietrowski, Solitar [1973] generalizing a deep result of Stallings [1971].

Theorem: A group G is the fundamental group of a finite graph of finite groups if and only if G has a finitely generated free group of finite index. □

Such groups G are called *free by finite.*

19. Corollary: If a group G has a finite pregroup structure, then G is free by finite. □

In section 7 we prove the converse result.

7. A Pregroup Structure for Graphs of Groups

In this section we construct a pregroup structure P for the fundamental group of a graph of groups $(\mathbf{H},Y)$. The pregroup P is such that the base groups $H_v \in \mathbf{H}$ correspond to a sequence of fundamental group systems for P. See definitions 6.4 and 6.5 for our non-standard notation regarding graphs of groups.

Fix a graph of groups $(\mathbf{H},Y)$ and a base $v_0 \in B$. Choose an orientation E^+ of E. Given $e \in E$, let $FP(e)$ be the free pregroup on e. Define

$$F = \bigoplus_{v \in B} H_v \bigoplus_{e \in E^+} FP(e)$$

to be the pregroup whose components are the base groups and the free pregroup on an orientation of the edges. We shall identify $\bar{e} \in \bar{E}^+$ with the element $e^{-1} \in F$. Set

$$P' = \bigcup_{v \in B} H_v \bigcup_{e \in E} \{ \, geh \in F \mid e \in E, \, g \in H_{\iota(e)}, \, h \in H_{\tau(e)} \, \}$$

$$D' = \{ \, (x',y') \in P' \times P' \mid x'y' \in P' \, \}$$

Eventually we shall see that (P',D') is a pregroup structure for $\mathbf{U}(F)$.

There is a relationship between P'-words and paths of Y which may be described as follows. Suppose $\alpha,\beta \in B \cup E$. We say that α *touches* β if one of the following conditions holds: $\alpha = \iota(\beta)$, $\beta = \tau(\alpha)$, $\alpha = \beta \in V$, or $\tau(\alpha) = \iota(\beta)$.

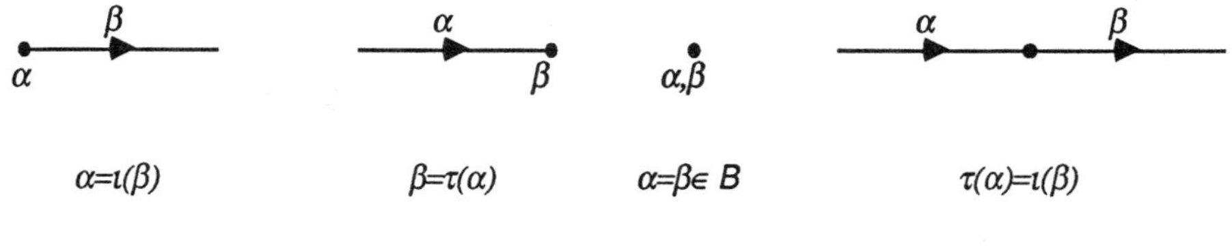

β α α β α,β α β

$\alpha = \iota(\beta)$ $\beta = \tau(\alpha)$ $\alpha = \beta \in B$ $\tau(\alpha) = \iota(\beta)$

Given a sequence $(\alpha_1, \ldots, \alpha_n) \in Y$, let $(\alpha_1, \ldots, \alpha_{i_1})$, $(\alpha_{i_1+1}, \ldots, \alpha_{i_2})$, $\ldots$, $(\alpha_{i_{(k-1)}+1}, \ldots, \alpha_{i_k} = \alpha_n)$ be the maximal subsequences generated by touching. Each such class $(\alpha_{i_{(j-1)}+1}, \ldots, \alpha_{i_j})$ determines a path p_j of the graph Y, and $(p_1, \ldots, p_k)$ is called the *path sequence of* $(\alpha_1, \ldots, \alpha_n)$. Given any element $x' \in P'$, we define $Y(x')$ to be an element of Y according to the formula

$Y(x') = e$ if $x' = g_1 e g_2$ for some $g_1 \in H_{\iota(e)}$, $g_2 \in H_{\tau(e)}$

$Y(x') = v$ if $x' \in G_v$

By the solution to the word problem in $\mathbf{U}(F)$, we see that $Y(x')$ is well defined for all $x' \in P'$. If $X' = (x_1', \ldots, x_n')$ is a P'-word, the *path sequence* $Y(X')$ *of* X' is defined to be the path sequence of $(Y(x_1'), \ldots, Y(x_n'))$. If the path sequence of X' is a single path p, then X' is called a p-word.

1. Lemma: Let $X'=(x_1', \ldots, x_n')$ be a P'-reduced word. Let $Z=(z_1, \ldots, z_m)$ be an F-reduced word representing $x_1'x_2' \cdots x_n' \in \mathbf{U}(F)$. Then the path sequence for X' equals the path sequence for Z, and Z is the unique F-reduced word representing $x_1' \cdots x_n'$.

Proof: Replace each element $x_i' \in X'$ with the F-reduced word representing x_i', obtaining an F-word $(v_1, \ldots, v_l)$. Clearly the path sequence is not disturbed by this operation. Now let $Z=(z_1, \ldots, z_m)$ be the F-reduced word obtained by performing elementary reductions on $(v_1, \ldots, v_l)$. Then Z is an F-reduced word representing $x_1'x_2' \cdots x_n'$ by definition. Theorem 1.4 implies Z is the unique F-reduced word representing $x_1' \cdots x_n'$. Since X' is P'-reduced, the process of elementary reductions cannot cancel edges, and so it follows that the path sequence for Z is the same as the path sequence for X'. □

Lemma 1 implies that P'-words representing the same element of $\mathbf{U}(F)$ have the same path sequence.

2. Lemma: Let $X'=(x_1', \ldots, x_n')$ be a P'-reduced p-word. If p is a path of length 0 or 1, then $l_{P'}(X') \leq 1$. If p is a path of length >1, then the edges of p are exactly $Y(X')=(Y(x_1'), \ldots, Y(x_n'))$.

Proof: The first claim is obvious. Suppose p is a path of length >1. If $Y(x_i) \in B$ for some $i<n$, then since $Y(x_i)$ touches $Y(x_{i+1})$, $x_i x_{i+1} \in P'$, contradicting the fact that X' is P'-reduced. If $Y(x_i) \in B$ for $i>1$ then $Y(x_{i-1})$ touches $Y(x_i)$, also a contradiction. Hence $Y(x_i) \in E$ for all $1 \leq i \leq n$, as desired. □

Suppose p is a path in Y, and X' and Y' are p-words. Evidently lemma 2 implies that $l_{P'}(X')= l_{P'}(Y')$.

3. Corollary: Suppose X' and Y' are P'-reduced words representing the same element of $\mathbf{U}(P')$. Then $l_{P'}(X')=l_{P'}(Y')$.

Proof: By lemma 1, X' and Y' have the same path sequence, say $(p_1, \ldots, p_k)$. Thus it suffices to consider the case where X' and Y' are P'-reduced p-words for some path p. By lemma 2, $l_{P'}(X')=l_{P'}(Y')$, since P'-length only depends on p.

4. Corollary: (Stallings) (P',D') is a pregroup, and $\mathbf{U}(P') \approx \mathbf{U}(F)$ in the natural way.

Proof: We verify hypotheses of theorem 1.3. By definition, (P',D') is closed under multiplication and inversion. Invariance of the length of reduced words follows from corollary 3. Since $F \subset P'$, P' generates $\mathbf{U}(F)$. □

Recall that given an edge $e \in E$, the edge group for e is E_e. By corollary 4, the following definition of $F(\mathbf{G},Y)$ is equivalent to that in Serre [1980, chap. 1.5.1]. Set

$$F(\mathbf{H},Y)=\mathbf{U}(P')/\ll ea^e\bar{e}=a^{\bar{e}} \mid e\in E, a\in E_e\gg.$$

We now construct a "quotient pregroup" P of P' and show that $\mathbf{U}(P)\approx F(\mathbf{H},Y)$.

Define an equivalence relation $\sim$ on P' by setting $x'\sim y'$ if $x'=y'\in H_v$ for some $v\in B$ or if $x'=geh$, $y'=ga^{\bar{e}}e(a^{-1})^e h$, for some $a\in E_e$. We define $P=P'/\sim$, and let $\rho:P'\to P$ denote the quotient map. The vertex groups embed naturally in P, and so are considered to be subsets of P. Notice that the notion of path sequence is compatible with $\sim$, so that we define the *path sequence of a P-word* $(x_1, \ldots, x_n)$ to be the path sequence of a representative P'-word $x_1' \cdots x_n'$.

We define $D\subset P\times P$ and $m:D\to P$ according to the following rules. Let $(x,y)\in P\times P$, and let $(x',y')\in P'\times P'$ represent (x,y). Then $(x,y)\in D$ if

(i) (x,y) is a p-word, where p is a path of length ≤ 1. In this case $xy=\rho(x'y')$.

or

(ii) (x,y) is a p-word, where p has edges $(e,\bar{e})$, and $x'y'=gea^e\bar{e}h$, for some $g,h\in H_{\iota(e)}$, $a\in H_e$. In this case $xy=ga^{\bar{e}}h\in G_{\iota(e)}$.

As an exercise, one may check that none of these these properties depend on the choice of representation, and that D and $m:D\to P$ are well defined.

5. Theorem: (Stallings) (P,D) is a pregroup.

Proof: Clearly (P,D) is closed under identity and inversion, so (P1) and (P2) hold. To verify (P3), first observe that an element of P is a p-word, where p is of length 0 or 1. Thus if $(w,x,y)_D$, we display the eight possible cases.

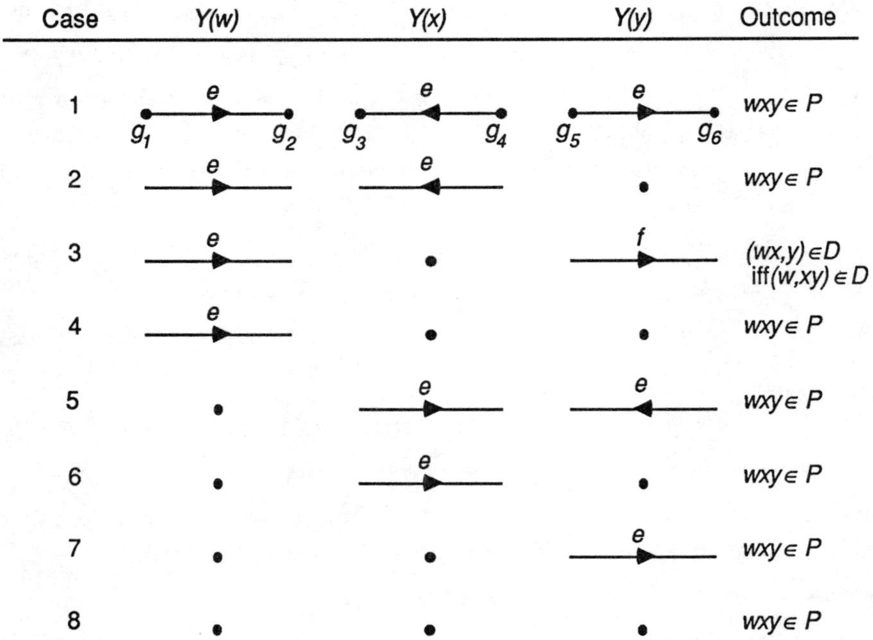

Case	$Y(w)$	$Y(x)$	$Y(y)$	Outcome
1	$g_1 \xrightarrow{e} g_2$	$g_3 \xleftarrow{e} g_4$	$g_5 \xrightarrow{e} g_6$	$wxy \in P$
2	$\xrightarrow{e}$	$\xleftarrow{e}$	$\bullet$	$wxy \in P$
3	$\xrightarrow{e}$	$\bullet$	$\xrightarrow{f}$	$(wx,y) \in D$ iff $(w,xy) \in D$
4	$\xrightarrow{e}$	$\bullet$	$\bullet$	$wxy \in P$
5	$\bullet$	$\xrightarrow{e}$	$\xleftarrow{e}$	$wxy \in P$
6	$\bullet$	$\xrightarrow{e}$	$\bullet$	$wxy \in P$
7	$\bullet$	$\bullet$	$\xrightarrow{e}$	$wxy \in P$
8	$\bullet$	$\bullet$	$\bullet$	$wxy \in P$

Case 1 may be checked as follows. Choose representatives $w'=g_1eg_2$, $x'=g_3\bar{e}g_4$, and $y'=g_5eg_6$. Then $(w,x)_D$ and $(x,y)_D$ imply the existence of $a \in E_e$ and $b \in E_{\bar{e}}=E_e$ such that $a^e=g_2g_3$ and $b^{\bar{e}}=g_4g_5$. Then

$$
\begin{aligned}
(w'x')y' &= (g_1a^{\bar{e}}g_4)g_5eg_6 \\
&= g_1a^{\bar{e}}b^{\bar{e}}eg_6 \\
&= g_1ea^eb^eg_6 \\
&= (g_1eg_2)(g_3b^eg_6) \\
&= w'(x'y')
\end{aligned}
$$

The other cases follow in a similar manner.

To verify (P4), suppose $(w,x,y,z)_D$. It suffices to check the possibility that $(Y(w),Y(x),Y(y))$ is as in case 3. This implies $(Y(x),Y(y),Y(z))$ is one of cases 5 to 8, and so $xyz \in P$. Thus $(w,x,y,z)_D$ implies $wxy \in P$ or $xyz \in P$, as desired. □

6. Lemma: Let τ:U$(P')\to$U$(P')/\{ ea^e\bar{e}=a^{\bar{e}} \} =F(\mathbf{H},Y)$ and $\rho:P'\to P$ be the quotient maps. Then τ factors through P as follows:

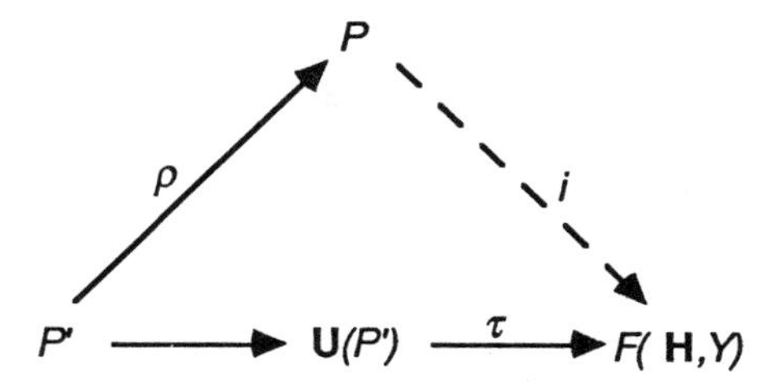

Moreover, $i:P\to F(\mathbf{H},Y)$ is a pregroup morphism.

Proof: Let $x\in P$, and let x' represent x. We define $i(x)=\tau(x')$. Notice that if $y'\in P'$ is another representative of x, then $y'=x'$ or $x'=geh$ and $y'=ga^{\bar{e}}e(a^{-1})^e h$. In the latter case

$$\tau(y')=\tau(g)\tau(a^{\bar{e}})\tau(e(a^{-1})^e h)=\tau(g)\tau(ea^e\bar{e})\tau(e(a^{-1})^e h)=\tau(geh)=\tau(x').$$

Thus $i:P\to F(\mathbf{H},Y)$ is well defined. Given $(x,y)\in D$ and representatives $x',y'\in P'$, we have $i(xy)=\tau(x'y')=\tau(x')\tau(y')=i(x)i(y)$, so i is a pregroup morphism. □

7. Theorem: (Stallings) The natural extension of i:U$(P)\to F(\mathbf{H},Y)$ by

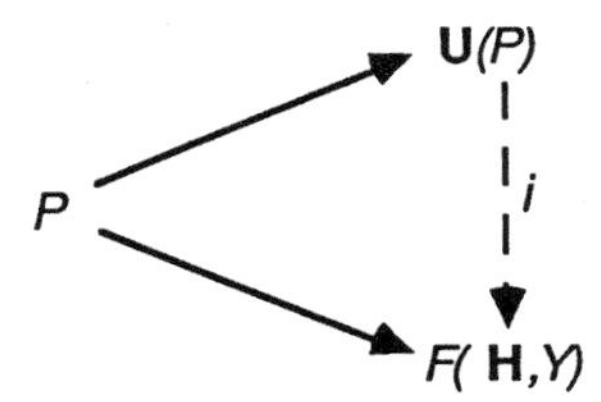

is an isomorphism.

Proof: The quotient map $\rho:(P',D')\to(P,D)$ is actually a pregroup morphism, so ρ extends naturally to ρ:U$(P')\to$U(P). We claim that $ker\,\tau\subset ker\,\rho$. Let $x'=ea^e\bar{e}(a^{-1})^{\bar{e}}$ be a generator for $ker\,\tau$. Note $(ea^e,\bar{e}(a^{-1})^{\bar{e}})$ is a P'-word so $\rho(x')=\rho(ea^e)\rho(\bar{e}(a^{-1})^{\bar{e}})=1$. Thus $ker\,\tau\subset ker\,\rho$, and so ρ factors through τ. Let $j:F(\mathbf{H},Y)\to$U(P) make the following diagram commute.

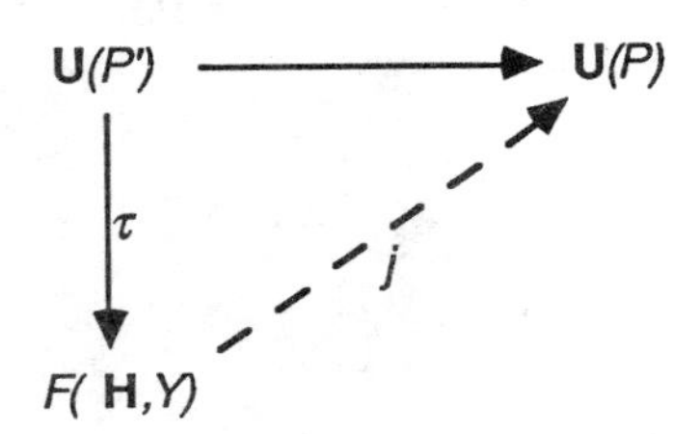

Now given $x \in P$, let x' represent x. Then $ji(x)=j\tau(x')=\rho(x')=x$. Since P generates $\mathbf{U}(P)$, $ji=1_{\mathbf{U}(P)}$. Given $x' \in P'$, notice $ij\tau(x')=i\rho(x')=\tau(x')$. Since $\tau(P')$ generates $F(\mathbf{H},Y)$, it follows that $ij=1_{F(\mathbf{H},Y)}$. Thus $i:\mathbf{U}(P)\approx F(\mathbf{H},Y)$ is an isomorphism. □

We identify the elements of $\mathbf{U}(P)$ and $F(\mathbf{H},Y)$ according to the isomorphism $i:\mathbf{U}(P)\rightarrow F(\mathbf{H},Y)$. The next step is to clarify the nature of the word problem in $F(\mathbf{H},Y)$.

8. Lemma: Let $x \in P$ and let $Y(x) \in E$. Then x is maximal in P.

Proof: Suppose $x \in P$ and $x \lesssim y$. Thus $y^{-1}x \in P$. Suppose $z \in P$ and $zx \in P$. If $Y(y) \in B$, then $Y(y)=\iota(Y(x))$ and hence $zy \in P$. Suppose $Y(y) \in E$. Then $Y(y)=Y(x)$. If $Y(z) \in B$, then evidently $zy \in P$ as desired. If $Y(z) \in E$, then $Y(z)^{-1}=Y(x)=Y(y)$. Let x_1ex_2, y_1ey_2, and $z_1\bar{e}z_2$ represent x, y, and z respectively. Then $y_1^{-1}x_1$ and z_2x_1 are in $\phi_{\bar{e}}(H_e)$. Hence $z_2y_1= z_2x_1(y_1^{-1}x_1)^{-1} \in \phi_{\bar{e}}(H_e)$, so $zy \in P$. We conclude that $y \lesssim x$, so x must be maximal in P.

9. Lemma: Let $X=(x_1, \ldots, x_n)$ and $Y=(y_1, \ldots, y_n)$ be P-reduced words representing the same element of $F(\mathbf{H},Y)$. Then X and Y have the same path sequence.

Proof: By theorem 1.4, X and Y differ by an interleaved product A. By lemma 3.2, the elements of A are all units. Consequently lemma 8 implies the elements of A are contained in the base groups. Thus interleaving by A does not alter the path sequence. □

Lemma 9 allows us to make the following definition.

10. Definition: Let $\alpha \in F(\mathbf{H},Y)$. Then the *path sequence* $Y(\alpha)$ of α is the path sequence of any reduced word representing α. □

Henceforth we fix a base $v_0 \in B$ and a maximal tree of Y. Let T be a maximal tree of Y with edges $E(T)$, and let X be the set of edges of Y that are not in T. A loop based at v_0 is a path in Y which begins and ends at v_0.

The following definitions of the fundamental groups $\pi_1(\mathbf{H},Y,v_0)$ and $\pi_1(\mathbf{H},Y,T)$ are exactly those of Serre [1980, chap. 1.5.1].

11. Definition: $\pi_1(\mathbf{H},Y,v_0)=\{\ \alpha\in F(\mathbf{H},Y)\ \mid\ Y(\alpha)$ is a loop based at $v_0\ \}$.

12. Definition: $\pi_1(\mathbf{H},Y,T)=F(\mathbf{H},Y)/\ll t=1\ \mid\ t\in E(T)\gg$. □

These two groups are canonically isomorphic, in that the quotient map $F(\mathbf{H},Y)\to\pi_1(\mathbf{H},Y,T)$ sends $\pi_1(\mathbf{H},Y,v_0)$ isomorphically to $\pi_1(\mathbf{H},Y,T)$. It is easy to see that definition 12 is equivalent to definition 6.6 for $\pi_1(\mathbf{H},Y,T)$ used in section 6.

Our strategy will be to find a subpregroup (Q,D_Q) of $F(\mathbf{H},Y)$ such that the inclusion $Q\subset F(\mathbf{H},Y)$ induces an isomorphism $\mathbf{U}(Q)\approx\pi_1(\mathbf{H},Y,v_0)$. We start by defining a pree (Q,D_Q) and verifying the pregroup axioms. We then investigate the order tree for Q. Once the order tree for Q is understood, an application of the proof of theorem A yields an isomorphism $\mathbf{U}(Q)\approx\pi_1(\mathbf{H},Y,T)\approx\pi_1(\mathbf{H},Y,v_0)$ as desired.

In order to define Q and to proceed with our investigation, we shall need some terminology. Given $v\in B$, let $t(v)$ be the geodesic from v_0 to v lying in the maximal tree. Let $d(v_0,v)$, the *distance* from v_0 to v, be the number of edges in $t(v)$. Let $p(v)$ be the round trip $t(v)t(v)^{-1}$. Given $x\in X$, let $v=\iota(x)$ and $w=\tau(x)$. Then $p(x)$ is the path $t(v)xt(w)^{-1}$.

We are now ready to define the pree (Q,D_Q). Eventually we shall see that (Q,D_Q) is a pregroup and $\mathbf{U}(Q)\approx\pi_1(\mathbf{H},Y,T)$.

13. Definition: The pregroup structure (Q,D_Q) for $\pi_1(\mathbf{H},Y,v_0)$ *induced by* T and v_0 is

$$Q=\{\ \alpha\in F(\mathbf{H},Y)\ \mid\ Y(\alpha)=p(z)\text{ for some }z\in X\cup B\ \}$$

$$D_Q=\ \}(\alpha,\beta)\in Q\times Q\ \mid\ \alpha\beta\in Q\ \}.\ \square$$

It is clear that (Q,D_Q) is a pree and it is also easy to verify pregroup axioms P1,P2, and P3. Thus the bulk of the work lies in proving axiom P4 and unraveling the order tree for Q. The next series of definitions and lemmas constitutes an investigation of the partial multiplication for D_Q. By this means we shall be able to prove Q is a pregroup.

14. Definition: Given $\alpha\in Q$, we say α is of *type* 0 if $Y(\alpha)=p(v)$ for some $v\in B$, and α is of *type* 1 if $Y(\alpha)=p(x)$ for some $x\in X$. A Q-word $(\alpha_1,\ldots,\alpha_n)$ is of type $(i_1,\ldots,i_n)$ if each α_j is of type $i_j\in\{\ 0,1\ \}$.

15. Lemma: Suppose $\alpha\in Q$. Then there exists a unique $z\in X\cup B$ such that $Y(\alpha)=p(z)$.

Proof: This follows immediately from lemma 9. □

16. Definition: Let $\alpha\in Q$ be of type 0. Let $v\in B$ be such that $Y(\alpha)=p(v)$. The element α is said to be *based at* v. Let l and r be elements of $F(\mathbf{H},Y)$ such that $Y(l)=Y(r^{-1})=t(v)$ and $\alpha=lr$. Then l is called a *left side* of α and r is called a right side of α. The set of left sides of α is denoted LS(α); the right sides RS(α).

17. Definition: Let $\alpha \in Q$ be of type 1. Let $x \in X$ be such that $Y(\alpha)=p(x)$. Then α is said to *cross over at* x. Let l, $r \in F(\mathbf{H},Y)$ be such that $Y(l)=t(\iota(x))$, $Y(r^{-1})=t(\tau(x))$, and $\alpha=lxr$. Then l and r are *left* and *right sides* of α, and the sets of left and right sides of α are denoted LS(α) and RS(α) respectively.

18. Definition: Let $\alpha \in Q$ be based at $v \in B$. Then v is also called the *left base* and the *right base* of α. If $\alpha \in Q$ crosses $x \in X$, then $\iota(x)$ is the *left base* of α and $\tau(x)$is the *right base* of α.

19. Definition: Let $v \in B$, and let $(t_1, \ldots, t_n)$ be the edges of $t(v)$. For $1 \leq i \leq n$, the path $(t_i, \ldots, t_n)$ is called a *left stub* of v and the path $(t_n^{-1}, \ldots, t_i^{-1})$ is called a *right stub* of v. The vertex v is both a *left* and a *right stub* of v.

20. Lemma: Let α, $\beta \in Q$. Let v be the right base of α and let w be the left base of β. Then (i) and (ii) are equivalent, and (iii) and (iv) are equivalent.

(i) there exists $r \in$ RS(α) and $l \in$ LS(β) such that $Y(rl)$ is a right stub of v.

(ii) for all $r \in$ RS(α) and $l \in$ LS(β), $Y(rl)$ is a right stub of v.

(iii) there exists $r \in$RS(α) and $l \in$LS(β) such that $Y(rl)$ is a left stub of w.

(iv) for all$r \in$RS(α) and $l \in$LS(β), $Y(rl)$ is a left stub of w.

Proof: By taking inverses, we see that (i $\Leftrightarrow$ ii) $\Leftrightarrow$ (iii $\Leftrightarrow$ iv). Clearly ii $\Rightarrow$ i. Now suppose that $r \in$RS(α), $l \in$LS(β) and rl is a right stub of v. If α is of type 1, say α crosses $x \in X$. If α is of type 0, set $x=1$. Let $l_\alpha \in$LS(α) be such that $l_\alpha xr=\alpha$. Let $r' \in$RS(α) and $l' \in$LS(β) be arbitrary. By the proof of lemma 9, there exists $g \in H_v$ and $h \in H_w$ such that $\alpha=l_\alpha xgr'$ and $\beta=l'hyr_\beta$. Hence

$l_\alpha xgr'l'hyr_\beta=\alpha\beta=l_\alpha xrlyr_\beta$, or $gr'l'h=rl$. Hence $r'l'$ and rl have the same path sequence, so that $Y(r'l')=Y(rl)$, a right stub of v. □

21. Definition: Let α, $\beta \in Q$, and let v be the right base of α and let w be the left base of β. Let $r \in$RS(α) and $l \in$LS(β). Then (α,β) *leans to the left* if rl is a right stub and (α,β) *leans to the right* if rl is a left stub of w.

22. Lemma: Suppose α, $\beta \in Q$ and (α,β) is of type (1,1). Then $\alpha\beta \in Q \Rightarrow$

(i) $\alpha\beta$ is of type 0,

(ii) there exists $x \in X$ such that α crosses at x and β crosses at x^{-1}, and

(iii) (α,β) leans to both the left and the right.

Proof: Suppose $\alpha\beta \in Q$. Let α cross at x and β cross at y. Be lemma 9, every element $z \in Q$ is such that $Y(z)$ has at most one edge not in $E(T)$. Now since (α,β) is of type (1,1), $Y(\alpha\beta)$ contains zero or two edges not in $E(T)$. Since $\alpha\beta \in Q$, we see that all the edges of $\alpha\beta$ are in $E(T)$, hence $\alpha\beta$ is of type 0. This proves (i) and (ii). Evidently, if $r \in$RS(α) and $l \in$LS(β), then $rl \in A_x$, for otherwise the x and the x^{-1}

will not cancel. This proves iii. □

23. Lemma: Suppose α, $\beta \in Q$ and $\alpha\beta \in Q$. Then

(i) (α,β) of type $(1,0)$ $\Rightarrow$ (α,β) leans to the left

(ii) (α,β) of type $(0,1)$ $\Rightarrow$ (α,β) leans to the right

(iii) (α,β) of type $(0,0)$ $\Rightarrow$ (α,β) leans to the left or to the right.

Proof: Clearly i $\Rightarrow$ ii upon taking inverses. Suppose (α,β) is of type $(1,0)$. Let α cross at x. Then evidently x is an edge of $Y(\alpha\beta)$, so $\alpha\beta$ also crosses at x. This implies that $Y(\beta)=p(v)$ for some $v \in t(\tau(x))$, and therefore if $r \in \mathrm{RS}(\alpha)$ and $l \in \mathrm{LS}(\beta)$, then $Y(rl)$ is the geodesic in the maximal tree from $\tau(x)$ to v_0. Hence (α,β) leans to the left. This establishes i and therefore also ii. To prove iii, suppose (α,β) is of type $(0,0)$. Then $\alpha\beta$ is of type 0. Let α be based at v, β at w, and $\alpha\beta$ at u. Evidently u is a vertex of $t(v)$ or of $t(w)$. This implies w is a vertex of $t(v)$ or v is a vertex of $t(w)$, respectively. In any event, $Y(rs)$ is the path in the maximal tree from v to w. Thus w is a vertex of $t(v)$ implies (α,β) leans to the left and v a vertex of $t(w)$ implies (α,β) leans to the right. □

24. Corollary: Suppose α, $\beta \in Q$ and $\alpha\beta \in Q$. Then (α,β) leans to the left or (α,β) leans to the right.

25. Lemma: Suppose α, β, $\gamma \in Q$ and $\alpha\beta$ and $\beta\gamma$ are in Q. Suppose $\beta\gamma$ leans to the left. Then $\alpha\beta\gamma$ is in Q if (β,γ) is of type $(0,0)$, $(1,0)$ or $(1,1)$.

Proof: Case 1: (β,γ) is of type $(0,0)$. Let β be based at w. Then there exists $l \in \mathrm{LS}(\beta)$ and $z \in F(\mathbf{H},Y)$, $Y(z)=t(w)^{-1}$, such that $\beta\gamma=lz$. By corollary 24, (α,β) leans to the left or to the right. Let $\alpha=l_\alpha x r_\alpha$ for $l_\alpha \in \mathrm{LS}(\alpha)$, $x \in X \cup \{1\}$ and $r_\alpha \in \mathrm{RS}(\alpha)$. Let v be the right base of α. Now $\alpha\beta\gamma=l_\alpha x(r_\alpha l)z$. Observe that

(i) if $x=1$, then (α,β) leans to the left $\Rightarrow$ $Y(\alpha\beta\gamma)=p(u)$ for some base u of $t(v)$;

(ii) if $x=1$, then (α,β) leans to the right $\Rightarrow$ $Y(\alpha\beta\gamma)=p(u)$ for some base u of $t(w)$;

(iii) if $x \in X$, then (α,β) leans to the left by lemma 22, and $Y(\alpha\beta\gamma)=p(x)$.

We conclude that $\alpha\beta\gamma \in Q$.

Case 2: (β,γ) is of type $(1,0)$ or $(1,1)$. Then lemma 22 and lemma 23 imply that (α,β) leans to the right and (β,γ) leans to the left, so the cancellations will proceed in an orderly manner. The results will be as follows.

(α,β,γ)	$Y(\alpha\beta\gamma)$
0,1,0	$p(x)$
0,1,1	$p(u)$ for some $u \in t(\iota(x))$
1,1,0	$p(u)$ for some $u \in t(\iota(x))$
1,1,1	$p(x)$ □

26. Remark: Notice $\alpha,\beta,\gamma \in Q$, $\alpha\beta$, $\beta\gamma \in Q$, and (β,γ) leans to the left *does not* imply $\alpha\beta\gamma \in Q$. For if (β,γ) is of type (0,1), and (α,β) is of type (0,0) and leans to the left, we could have a situation such as $\alpha=t_1t_2gt_2^{-1}t_1^{-1}$, $\beta=t_1ht_1^{-1}$, $\gamma=t_1h^{-1}xt_3$. Thus $Y(\alpha\beta\gamma) \neq p(x)$, so $\alpha\beta\gamma \notin Q$.

27. Corollary: The pree (Q,D) is indeed a pregroup.

Proof: Clearly Q satisfies P1 and P2. As for $P3$, if $(\alpha,\beta,\gamma)_{D_Q}$, then $(\alpha\beta,\gamma)_{D_Q} \Leftrightarrow (\alpha\beta)\gamma \in Q \Leftrightarrow \alpha(\beta\gamma) \in Q \Leftrightarrow (\alpha,\beta\gamma)_{D_Q}$. Now suppose that $(\alpha,\beta,\gamma,\delta)_{D_Q}$. By corollary 24, we may assume (β,γ) leans to the left, taking inverses if necessary. Let $(\alpha,\beta,\gamma,\delta)$ be of type (i_1,i_2,i_3,i_4), where each $i_j \in (0,1)$. If $(i_2,i_3) \neq (0,1)$ then lemma 25 implies $\alpha\beta\gamma \in Q$. If $(i_2,i_3)=$ (0,1), then (γ^{-1},β^{-1}) leans to the left by lemma 23. Hence lemma 25 implies $\delta^{-1}\gamma^{-1}\beta^{-1} \in Q$ and therefore $\beta\gamma\delta \in Q$. Thus Q satisfies P4. □

Now that we have shown Q is a subpregroup of $\pi_1(\mathbf{H},Y,v_0)$, we can apply the general theory of the earlier sections to Q. In order to succeed in this endeavor, we must make two assumptions. The first of these assumptions concerns the graph Y and is of fundamental importance. The second assumption, concerning $(\mathbf{H},Y)$, is largely a matter of convenience.

28. Definition: Y is said to be of finite diameter if the set $\{\ d(v,v_0) \in \mathbb{Z} \mid v \in B\ \}$ is bounded above.

Assumption 1: The graph Y is of finite diameter.

29. Definition: $(\mathbf{H},Y)$ is said to be *proper* if for each edge $t \in E(T)$, the edge maps $\alpha{:}H_t \to H_{\iota(t)}$ and $\beta{:}H_t \to H_{\tau(t)}$ are *not* onto.

Assumption 2: The graph Y is proper.

Suppose Y satisfies assumption 1 but not assumption 2. Let T be a maximal tree and v_0 a basepoint for Y. Let Q be the induced pregroup for $\pi_1(\mathbf{H},Y,v_0)$. Say an edge of T is degenerate if one of its edge maps is onto. By systematically removing the degenerate edges of Y, we obtain a graph of groups $(\hat{\mathbf{H}},\hat{Y})$ with maximal tree $\hat{T}$,

basepoint $\hat{v}_0$ and induced pregoup $\hat{Q}$ for $\pi_1(\hat{\mathbf{H}},\hat{Y},\hat{v}_0)$ such that

(i) there exists an isomorphism $\phi:\pi_1(\mathbf{H},Y,v_0)\rightarrow\pi_1(\hat{\mathbf{H}},\hat{Y},\hat{v}_0)$, and

(ii) $\phi|_Q:Q\rightarrow\hat{Q}$ is a pregroup isomorphism.

In particular, $\hat{Q}$" is a pregroup structure for $\pi_1(\hat{\mathbf{H}},\hat{Y},v_0)$ $\Leftrightarrow$ Q is a pregroup structure for $\pi_1(\mathbf{H},Y,v_0)$.

There is one subtle point. Without assumption 1, we may be unable to contract $(\mathbf{H},Y)$ to a proper graph of groups $(\hat{\mathbf{H}},\hat{Y})$. For example, if F_i is the free group on i generators, and $F_i\subset F_{i+1}$ is a monomorphism, then

$$F_1 \xrightarrow{F_1} F_2 \xrightarrow{F_2} F_3 \xrightarrow{F_3} \bullet \bullet \bullet$$

cannot be contracted. It is also worth noting that a proper graph Y still may have degenerate edges, provided they are not in the maximal tree. We shall deal with this point later. Notice we do not wish to contract such an edge, because then we lose control over which edges are in the maximal tree.

Henceforth we shall assume $(\mathbf{H},Y)$ is a proper graph of groups with finite diameter, edges E, bases B, maximal tree T with edges $E(T)$, basepoint v_0, and induced pregroup Q. The first step is to work out the relationship between T and the tree ordering $\precsim$ of the pregroup Q.

30. Lemma: Suppose $\alpha\in Q$ is of type 1. Then α is maximal in Q.

Proof: Suppose $\beta\in Q$ and $\alpha\precsim\beta$. Let $\gamma\in Q$, and suppose that $\gamma\alpha\in Q$. We claim that $\gamma\beta\in Q$. Notice $\alpha^{-1}\beta\in Q$, hence $(\alpha,\alpha^{-1}\beta)$ leans to the left by lemmas 22 and 23. By lemma 25, $\gamma\beta=\gamma\alpha(\alpha^{-1}\beta)\in Q$. We conclude that $\beta\precsim\alpha$, so $\alpha\approx\beta$. It follows that α is maximal in Q. □

31. Corollary: Suppose α is a unit of Q. Then α is of type 0.

32. Lemma: Suppose α, $\beta\in Q$ and $\alpha<\beta$. Let v be the base of α and w be the left base of β. Then v is a base of $t(w)$. If $v=w$, then β is of type 1.

Proof: Notice corollary 31 implies α is of type 0. Hence v is also the base of α^{-1}. Thus either v is a base of $t(w)$ or w is a base of $t(v)$. Suppose w is a base of $t(v)$. We claim that $w=v$ and β is of type 1.

Case 1: $w\neq v$. Then $\alpha^{-1}\beta\in Q$ implies β is of type 0 and (α^{-1},β) leans to the left. Hence $(\alpha,\alpha^{-1}\beta)$ leans to the left. Since $(\alpha,\alpha^{-1}\beta)$ is of type (0,0), lemma 25 implies $\beta\precsim\alpha$, a contradiction.

Case 2: $w=v$. If β is of type 0, the argument of case 1 applies, yielding a contradiction. Hence β is of type 1.

33. Corollary: Q is of finite height.

Proof: Suppose we have an infinite chain of strict inequalities $\alpha_1<\alpha_2<\cdots$. By corollary 31, each α_i is of type 0. Hence lemma 32 implies the bases $v_0,v_1,\ldots,v_i,\cdots$ of each α_i are distinct bases of some geodesic in Y. This contradicts the fact that Y is of finite diameter. □

By corollary 33, the depth of an element of Q is well defined.

34. Lemma: Suppose $\alpha,\beta\in Q$ are of type 0. Suppose that for some m, $0\leq m\leq d$, we have $\alpha\approx_m\beta$. Let v be the base of α, and let w be the base of β. Then $v=w$.

Proof: We know that $\alpha^{-1}\beta\in Q$, so that v is a vertex of $t(w)$ or w is a vertex of $t(v)$. We may assume v is a vertex of $t(w)$ and $v\neq w$. Let

(i) $(t_1,\ldots,t_m)$ be the edges of $t(v)$, and

(ii) $(t_1,\ldots,t_m,\ldots,t_n)$ be the edges of $t(w)$. Since $(\mathbf{H},Y)$ is proper, there exists $h\in H_v$ such that $t_{m+1}^{-1}ht_{m+1}\notin H_{\tau(t_{m+1})}$. Choose $l_\alpha\in \mathrm{LS}(\alpha)$, $r_\alpha\in \mathrm{RS}(\alpha)$, $l_\beta\in \mathrm{LS}(\beta)$ and $r_\beta\in \mathrm{RS}(\beta)$ such that $\alpha=l_\alpha r_\alpha$ and $\beta=l_\beta r_\beta$. Then $l_\alpha^{-1}l_\beta$ is a left stub of β, say $\hat{h}\in H_v$ and $s\in F(\mathbf{H},Y)$ satisfy

$$l_\alpha^{-1}l_\beta=\hat{h}t_{m+1}h_{m+1}\cdots t_nh_n=\hat{h}t_{m+1}s.$$

It follows that $\beta=l_\alpha\hat{h}t_{m+1}sr_\beta$.

We need a special element $\gamma\in U^i(Q)$. Let $\gamma=l_\beta s^{-1}t_{m+1}^{-1}h\hat{h}^{-1}l_\alpha^{-1}$. Thus $Y(\gamma)=p(w)$, so that $\gamma\in Q$. Moreover, $\gamma^{-1}\beta=l_\alpha\hat{h}ht_{m+1}sr_\beta\in Q$, so γ^{-1} and β are comparable in Q. Since β and γ are of type 0, lemma 25 implies $\gamma\approx\beta$, so $\gamma\approx_m\beta$ by lemma 3.6. In particular, $\gamma\in U^i(Q)$. Notice $\gamma\alpha=l_\beta s^{-1}t_{m+1}^{-1}h\hat{h}^{-1}r_\alpha\in Q$, but $\gamma\beta=l_\beta s^{-1}t_{m+1}^{-1}ht_m sr_\beta$. Since $t_{m+1}^{-1}ht_{m+1}\notin H_{\tau(t_{m+1})}$, the cancellations are obstructed and $\gamma\beta\notin Q$. Thus β is *not* $\lesssim_m\alpha$, contradicting the hypothesis $\alpha\approx_m\beta$.

35. Corollary: Suppose α, $\beta\in Q$ and α and β are of type 0. Suppose $\alpha^{-1}\beta\in Q$. Let α be based at v and β be based at w. Let m be the depth of β. Then

(i) v is a base of $t(w)$ and $v\neq w \Leftrightarrow \alpha<_m\beta$

(ii) $v=w\Leftrightarrow \alpha\approx_m\beta$.

Proof: Use lemmas 32, 34, 3.6, and 3.8. □

Henceforth, we shall let d denote the depth of Q. Since Q is of finite height, the general theory shows that each $U^i(Q)$ has a generating system $(\mathbf{G}_i,X_i)$. Obviously the elements of $\mathbf{G}_i$ should have something to do with the base groups H_v. Hence the following definition.

36. Definition: Let $v \in B$. Then the group G_v is defined as follows. Let $(t_1, \ldots, t_n)$ be the edges of $t(v)$ and let $l = t_1 \cdots t_n \in F(\mathbf{G},Y))$. Then $G_v = \{ l^{-1}hl \mid h \in H_v \}$. □

Clearly each $G_v \subset Q$, and in fact the G_v's enjoy many special properties. Because $(\mathbf{H},Y)$ is proper, for all $v \in B$ there exists $g_v \in G_v$ such that g_v is based at v. In view of corollary 35, we may make the following definition.

37. Definition: Let $v \in B$. The depth of v is the depth of g_v, where $g_v \in G_v$ is such that g_v is based at v. □

Evidently v_0 has depth d, and corollary 35 implies $U^d(P) \subset G_{v_0}$. Indeed, the only elements α of Q such that $Y(\alpha) = v_0$ are the elements of H_{v_0}, and $G_{v_0} = H_{v_0}$ by definition.

38. Corollary: $U^d(P) = G_{v_0}$. □

In fact, each G_v is the fundamental group of some maximal vertex V in $U^m(Q)$ where m is the depth of v.

39. Lemma: Let $v \in B$ be of depth m. Let $g_v \in G_v$ be based at v. Let V be the $U^m(Q)$-maximal vertex containing g_v. Then $G_v = G_V$, the fundamental group of V.

Proof: By corollary 35, $G_v \subset U^m(Q)$. Thus the maximality criterion for G_V expressed in theorem 4.10 implies that $G_v \subset G_V$. Now suppose that $g \in G_V$. By corollary 35, g is based at some $w \in B$ such that w is a base of $t(v)$. Let $(t_1, \ldots, t_k)$ be the edges of $t(w)$ and $(t_1, \ldots, t_k, \ldots, t_n)$ be the edges of $t(v)$. Then gg_v and $g_v g$ are in Q, so that $g = t_1 \cdots t_k h t_k^{-1} \cdots t_1$ for some $h \in H_w$. If g is maximal in $U^m(Q)$, then corollary 35 implies that $w = v$, so $h \in H_v$ and therefore $g \in G_v$. If g is a unit in $U^m(Q)$, then g stabilizes V. It follows that $gg_v \approx_i g_v$, so $g_v^{-1} g g_v \in Q$. Thus $t_n^{-1} \cdots t_{k+1}^{-1} h t_{k+1} \cdots t_n \in H_v$, say $t_n^{-1} \cdots t_{k+1}^{-1} h t_{k+1} \cdots t_n = \hat{h} \in H_v$. Then $h = t_{k+1} \cdots t_n \hat{h} t_n^{-1} \cdots t_{k+1}^{-1}$, and so it follows that $g = t_1 \cdots t_n \hat{h} t_n^{-1} \cdots t_1^{-1} \in G_v$. Thus $G_V \subset G_v$, as desired. □

Let $v \in B$ and let m be the depth of v. Let V be a $U^m(Q)$ maximal vertex. If $G_v = G_V$, the v and V are said to *correspond.* We see that each G_v is the fundamental group of some cyclic $U^m(Q)$-maximal vertex, where m is the depth of v. Hence it is reasonable to suppose the G_v's form the fundamental group systems for the various $U^i(Q)$'s. This statement is almost correct. Indeed, for $i > 0$, each $\simeq_i$-class contains exactly one orbit O, and this orbit contains a vertex V such that $G_V = G_v$ for some $v \in B$. Moreover, if v and w are distinct bases of equal depth m, then the corresponding vertices V and W lie in distinct $\simeq_m$-classes. From this follows that $\{ G_v \mid \text{depth of } v \text{ is } i \}$ is a fundamental group system for each $i > 0$, and the spanning set for $U^i(Q)$ is empty.

The situation at level 0 is more complicated, because this is where the type 1 elements live. We shall see that each $\simeq$-class contains two or one orbits, depending on how and where the HNN'ing occurs. The next series of lemmas will validate all these assertions.

40. Lemma: Let $\alpha \in Q$ be of type 0, say α is based at $v \in B$. Let α be of depth $m<d$. Then there exist units u and w of $U^m(Q)$ such that $u\alpha w \in G_v$.

Proof: Let $\alpha = h_0 t_1 h_1 \cdots t_n h_n t_n^{-1} \hat{h}_{n-1} \cdots t_1^{-1} \hat{h}_0$. Let $u = t_1 t_2 \cdots t_{n-1} h_{n-1}^{-1} t_{n-1} \cdots h_1^{-1} t_1 h_0^{-1}$ and similarly let $w = \hat{h}_0^{-1} t_1 \cdots t_{n-1} \hat{h}_{n-1} t_{n-1}^{-1} t_{n-2}^{-1} \cdots t_1^{-1}$. Then u and v are units of $U^m(Q)$ by corollary 35. Clearly $u\alpha w \in G_v$. □

41. Lemma: Let $\alpha \in Q$ be cyclic of type 0, say α of depth $m<d$ and α based at $v \in B$. Then there exists a unit u of $U^m(Q)$ such that $u^{-1}\alpha u \in G_v$.

Proof: Choose u and w units of $U^m(Q)$ such that $u^{-1}\alpha w = \beta \in G_v$. Then $\alpha = u\beta w^{-1}$. Since $\alpha^2 \in Q$, $u\beta(w^{-1}u)\beta w^{-1} = \alpha^2 \in Q$. By construction of u and w in lemma 40 we see that β is based at v. It follows that $w^{-1}u = g \in G_v$. Thus $u^{-1}\alpha u = \beta g \in G_v$, as desired. □

42. Lemma: Let V be a $U^m(Q)$-maximal vertex for some $m>0$. Then V is cyclic. □

Proof: Since $m > 0$, corollary 31 implies V contains an element α of type 0. If $m=d$, use corollary 38. Suppose $m > d$. Let $\alpha = h_0 t_1 \cdots t_n h_n t_n^{-1} \hat{h}_{n-1} \cdots t_1^{-1} \hat{h}_0$. Let $u = \hat{h}_0^{-1} t_1 \hat{h}_1^{-1} \cdots t_{n-1} \hat{h}_{n-1}^{-1} h_{n-1}^{-1} t_{n-1}^{-1} h_{n-2}^{-1} \cdots t_1^{-1} h_0^{-1}$. Then u is a unit of $U^m(Q)$ by corollary 35. Moreover, $\alpha u \in Q$, so $\alpha \approx_i \alpha u$, and clearly $\alpha^2 \in Q$. Thus V is cyclic.

43. Lemma: Let $i \in \mathbb{Z}$ be such that $1 \le i \le d$. Let

$$\mathbf{B}_i = \{\ V \subset U^i(Q) \mid V \text{ is a } U^i(Q)\text{-maximal vertex corresponding to a base } b \text{ of depth } i\ \}.$$

Then $\mathbf{B}_i$ is a basepoint set for $U^i(Q)$. Let $\mathbf{G}_i = \{\ G_v \mid v \in B \text{ and } v \text{ is of depth } i\ \}$. Then $\mathbf{G}_i$ is the fundamental group system for $U^i(Q)$ based at $\mathbf{B}_i$, and $o_i = (\mathbf{G}_i, \varnothing)$ is a generating system for $U^i(Q)$. Moreover, $(o_1, \ldots, o_d)$ is a good $U(Q)$-sequence.

Proof: By corollary 38, we may assume $i < d$. Let $\tilde{\mathbf{B}}_i$ be any basepoint system for $U^i(Q)$. Let $\tilde{V} \in \tilde{\mathbf{B}}_i$. By lemma 42, $\tilde{V}$ is cyclic. Let $\tilde{g} \in G_{\tilde{V}} \cap \tilde{V}$. By lemma 41, there exists $u \in U^{i+1}(Q)$ such that $u\tilde{g}u^{-1} \in G_v$, for some $v \in B$ of depth m. Let V be the $U^i(Q)$-maximal vertex associated with V. Evidently $V = u\tilde{V}$, so we may replace $\tilde{V}$ by V. We conclude that there exists a basepoint set $\mathbf{B}_i$ for $U^i(Q)$ such that for each $V \in \mathbf{B}_i$, $G_V = G_v$ for some v of depth i.

Now suppose that v and w are distinct bases of depth i in the same $\simeq_i$-class. Let V and W be the associated vertices. Then there exists $\alpha \in V$ such that $\alpha^{-1} \in W$. By lemma 40, choose u and $w \in U^{i+1}(Q)$ such that $u\alpha w$ is cyclic. Then $u\alpha w \approx_i w^{-1}\alpha^{-1}u^{-1}$. However, $\alpha \approx_i \alpha w$ and $\alpha^{-1} \approx_i \alpha^{-1}u^{-1}$, so that α and α^{-1} are in the same orbit, a contradiction.

We conclude that

(i) each $v \in B$ of depth i corresponds to a vertex $V \in \mathbf{B}_i$,

(ii) this correspondence is 1-1,

(iii) $\mathbf{G}_i$ is the fundamental group system for $U^i(Q)$ based at $\mathbf{B}_i$, and

(iv) the spanning sets are empty, since each $\simeq_i$-class contains exactly on orbit.

Thus $(\mathbf{G}_i, \varnothing)$ is a generating system for $U^i(Q)$.

For each i, $i=1, \ldots, d$, let $o_i = (\mathbf{G}_i, \varnothing)$. Clearly $(o_d, \ldots, o_1)$ is a good $U(Q)$-sequence. □

We proceed to describe the generating set for Q.

44. Definition: Henceforth $\mathbf{E}_A$ will denote the set of $\simeq$-classes which *do not* contain an element of type 1, and $\mathbf{E}_H$ will be the set of $\simeq$-classes which contain an element of type 1.

45. Definition: Let $v \in B$ be a base. Suppose $t(v)$ is not a proper subpath of any other geodesic in T based at v_0. Then v is called an *extreme base*. An extreme base is said to be an *amalgam base* if v is not the initial or terminal base of any edge $x \in X$. The set of amalgam bases is denoted am.

46. Lemma: Each $\simeq$-class $E \in \mathbf{E}_A$ contains exactly one orbit O. There is a set $\mathbf{B}_A$ of Q-maximal vertices representing these orbits such that each $V \in \mathbf{B}_A$ corresponds to a base $v \in \text{am}$. Moreover, this correspondence is 1-1.

Proof: The above analysis applies here, since all the elements involved are of type 0. □

47. Notation: Let $x \in X$ have left base v and right base w. Let $(t_1, \ldots, t_m)$ be the edges of $t(v)$ and $(\hat{t}_1, \ldots, \hat{t}_n)$ be the edges of w. Then $l_x = t_1 \cdots t_m$ is the *left side* of x and $r_x = \hat{t}_n^{-1} \cdots \hat{t}_1^{-1}$ is the *right side* of x. Recall that E_x is the edge group of x. Let $\alpha(E_x)$ be the image of E_x in H_v. We define $\hat{E}_x = l_x \alpha(E_x) l_x^{-1} \subset G_v$. □

Given $x \in X$, $\alpha = l_x x r_x \in Q$ is a simple maximal element of Q, so we may speak of the attaching group A_α of α. Recall from remark 5.2 that if V is the Q-maximal vertex containing α, then

$$A_\alpha = G_V \cap U(Q) = \{ \, a \in U(Q) \mid (\alpha^{-1}, a, \alpha) \text{ associates } \}.$$

48. Lemma: Let $x \in X$, and let v be the right base of x. Let $\alpha = l_x x r_x$. Let V be the Q-maximal vertex containing α. Then

(i) If $V = v_0$, then V is simple, $\alpha = x$, and $G_V = A_x = \hat{E}_x$.

(ii) If $V \neq v_0$ and V is simple, then $G_V = A_\alpha = \hat{E}_x$.

(iii) If $V \neq v_0$ and V is cyclic, the situation is as follows. Let $(t_1, \ldots, t_n)$ be the edges of $t(v)$. Let $w = \iota(t_n)$. Then $G_V = G_v$ and $A_\alpha = G_v \cap G_w$.

Proof: (i) Suppose $v = v_0$. We see that $\alpha = x$. Let $\beta \in V$, so that $\beta \approx x$. Since $x^{-1}\beta \in Q$, either $\beta = G_v = H_{v_0}$, or for some $h \in H_{v_0}$ and $a \in A_x$, $\beta = axh$. Suppose $\beta \in H_{v_0}$. Then $x\beta \in Q$, contradicting the facts that $x \lesssim \beta$ and $x^2 \notin Q$. Thus the other alternative holds, and V is simple. From the remark preceding the lemma we deduce that $G_V = A_x = \hat{E}_x$.

(ii) Notice $a \in A_\alpha \Leftrightarrow \alpha^{-1} a \alpha \in Q \Leftrightarrow a = \hat{E}_x$.

(iii) Let $g \in V$ be cyclic. Then g is of type 0. Let g be based at $w \in B$. Then $g^{-1}\alpha \in Q \Rightarrow w$ is a base of $t(v)$. By corollary 35, $w = v$. Thus V corresponds to v and $G_V = G_v$. In this case $A_\alpha = G_V \cap U = G_v \cap U = G_v \cap G_w$. □

We now investigate the circumstances under which conditions (ii) and (iii) of lemma 48 will hold. Given $x \in X$, say x is *degenerate on the left* if $\hat{E}_x = G_v$.

49: Lemma: Let V be a maximal vertex of Q containing an element $\alpha \in l_x x r_x$ for some $x \in X$. Let v be the left base of α. Then V is cyclic $\Leftrightarrow$ { x is degenerate on the left, depth $(v) = 0$, and for all $y \in X$, $\iota(y) = v \Rightarrow y = x$ }.

Proof: ($\Rightarrow$) By corollary 38, $v \neq v_0$. Thus lemma 48 implies $G_V = G_v$. Evidently depth$(v) = 0$. Notice $\hat{E}_x \subset G_v$ by definition. Suppose that $g \in G_v$. Then $g \in G_V$, so lemma 4.14a implies $\alpha^{-1} g \alpha \in Q$. From this we deduce that $g \in \hat{E}_x$. Thus $\hat{E}_x = G_v$, so x is degenerate on the left. Finally, suppose that $y \in X$ is such that $\iota(y) = x$. Choose $g_v \in G_v$ such that g_v is based at v. By lemma 39, $g_v \approx \alpha$. Notice $l_y = l_x$, so $r_y^{-1} y^{-1} l_x^{-1} g_v \in Q$. It follows that $r_y^{-1} y^{-1} l_y^{-1} \alpha = r_y^{-1} y^{-1} x r_x \in Q$, so $y = x$. ($\leftarrow$) Conversely, suppose $\hat{E}_x = G_v$, depth$(v) = 0$, and for all $y \in X$, $\iota(y) = v \Rightarrow y = x$. Choose $g_v \in G_v$ based at v. Suppose $\beta \in Q$ is such that $\beta g_v \in Q$. The hypotheses imply that either β is of type 0 based at a vertex w in $t(v)$, or else β crosses at x^{-1}. If β is of type 0, then clearly $\beta\alpha \in Q$, and if β crosses at x^{-1}, then $\hat{E}_x = G_v$ implies $\beta\alpha \in Q$. Thus $l_x x r_x \lesssim g_v$, and clearly $g_v \lesssim l_x x r_x$. We conclude that $g_v \in V$, so V is cyclic. □

50. Lemma: Let E be a $\simeq$-class of Q and suppose that E has more than one orbit. Then there exists a unique $x \in X$ such that $\alpha = l_x x r_x \in E$. Moreover, E has two orbits, one containing α and the other containing α^{-1}.

Proof: By lemma 46, E contains an element β of type 1, say β crosses at x. Using the argument of lemma 40, we see that $\alpha = l_x x r_x$ is in the same orbit as β. Let O_x be the orbit of E containing α. Suppose $\gamma \in O_x$ is such that γ^{-1} is in the orbit $O \subset E$, and $O_x \neq O$. As in the proof of lemma 43, we see that γ is of type 1, say γ crosses at y. Since γ and α are in the same orbit, there exists units u and w of Q such that $u\gamma w = \alpha$. Let $O_{x^{-1}}$ be the orbit containing α^{-1}. We have shown that O and $O_{x^{-1}}$ are the only orbits of E, and that $O \neq O_{x^{-1}}$, as desired. □

Henceforth we fix a orientation X^+ of X.

50a. Definition: Let $\mathbf{I} = \{ V_{x^{-1}} \subset Q \mid V_{x^{-1}} \text{ is cyclic and } x \in X^+ \}$ be called the set of *improper HNN basepoints*. Let $\mathbf{P} \doteq \{ V_x \in Q \mid V_{x^{-1}} \text{ is simple and } x \in X^+ \}$ be the set of *proper* HNN basepoints.

50b. Lemma:. Let

$$\begin{aligned} \mathbf{G}_0 = \{ G_v \mid v \in \text{am} \} \quad & \cup \{ G_v \mid v = \tau(x), x \in X^+, V_{x^{-1}} \in \mathbf{I} \} \\ & \cup \{ \hat{E}_{x^{-1}} \mid x \in X^+, V_{x^{-1}} \in \mathbf{P} \}. \end{aligned}$$

Then $(\mathbf{G}_0, \tilde{X})$ is the generating system for Q based at $\mathbf{B}$.

Proof: Use lemmas 46, 49, and 50.

The following result summarizes our analysis.

Theorem B: Let $(\mathbf{H}, Y)$ be a graph of groups with bases B, edges E, basepoint v_0, and maximal tree T. Suppose the graph of Y is of finite diameter, and $(\mathbf{H}, Y)$ is proper. Let $X \subset E$ be the edges of Y not in T, and let X^+ be an orientation of X. Then $\pi_1(\mathbf{H}, Y, v_0)$ has a pregroup structure Q satisfying the following conditions:

(i) for some $d \geq 0$, Q has depth d, and a good Q-sequence $((\mathbf{G}_d, \varnothing), \ldots, (\mathbf{G}_1, \varnothing), (\mathbf{G}_0, \tilde{X}))$ satisfying (ii), (iii), and (iv):

(ii) for $i = 1, \ldots, d$, the groups of $\mathbf{G}_i$ are in 1-1 correspondence with the maximal bases of depth i, and corresponding fundamental groups and base groups are isomorphic,

(iii) $\tilde{X}^+$ is in 1-1 correspondence with X^+,

(iv) $\mathbf{G}_0$ is in 1-1 correspondence with

$\{ v \in B \mid \text{depth}(v) = 0 \} \cup \hat{B}$, where $\hat{B} \subset \{ v \in B \mid \text{depth}(v) > 0 \}$.

If $G \in \mathbf{G}_0$ corresponds to G_v for some $v \in B$ of depth 0, then G is isomorphic to $H_v \in \mathbf{H}$. If $G \in \mathbf{G}_0$ corresponds to some $v \in \hat{B}$, then G is isomorphic to a subgroup of H_v.

Proof: Let $(o_1, \ldots, o_d)$ be as in lemma 43 and o_0 as in lemma 50b. Then $(o_0, \ldots, o_d)$ is a good Q-sequence satisfying properties (i), (ii), (iii), and (iv). We now use the proof of theorem A to construct a graph of groups $(\hat{\mathbf{H}},\hat{Y})$ with maximal tree $\hat{T}$ such that $\mathbf{U}(Q)\approx\pi_1(\hat{\mathbf{H}},\hat{Y},\hat{T})$. We investigate the structure of $(\hat{\mathbf{H}},\hat{Y})$. The resulting graph $\hat{Y}$ is somewhat different from Y, but nevertheless, $\pi_1(\hat{\mathbf{H}},\hat{Y},\hat{T})\approx\pi_1(\mathbf{H},Y,T)$ in such a way that the diagram

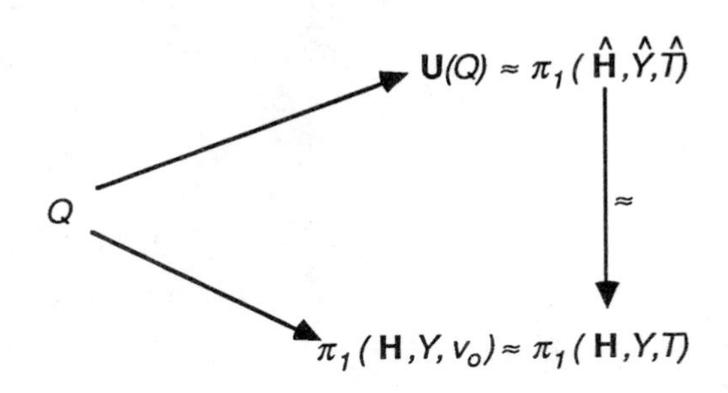

commutes. We conclude that Q is a pregroup structure for $\pi_1(\mathbf{H},Y,T)$. □

51: Corollary: A group G is free by finite if and only if G is the universal group of a finite pregroup. □

We conclude this section with the following examples. These should be compared with examples 3.A.5.4 and 3.A.5.5 in Stallings [1971].

52. Example: Let $(\mathbf{H},Y)$ be the graph of groups

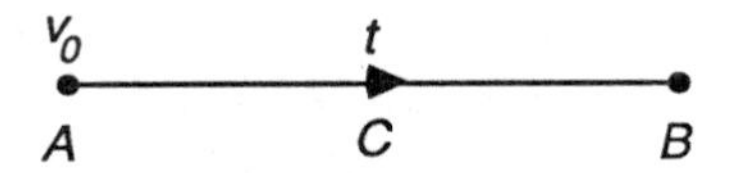

so $\pi_1(\mathbf{H},Y)=A*_C B$. Then the pregroup structure for $\pi_1(\mathbf{H},Y)$ is

$$\{\ (atbt^{-1}a')^{\pm 1}\in F(\mathbf{H},Y)\ \mid\ a,a'\in A \text{ and } b\in B\ \}.$$

53. Example: Let $(\mathbf{H},Y)$ be the graph

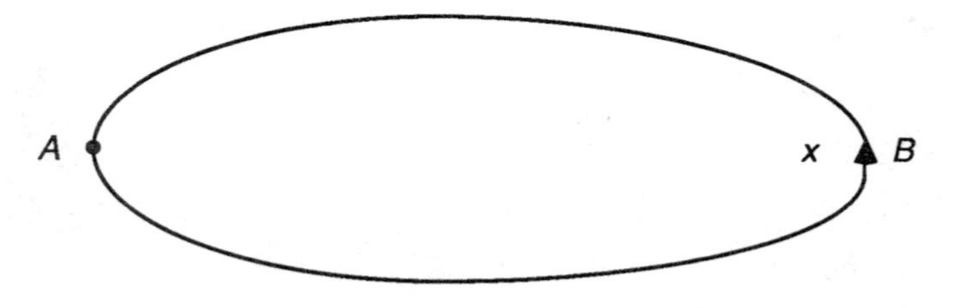

Then the pregroup structure for the HNN group $A \circlearrowleft B$ is $\{\ (ata')^{\pm 1}\in F(\mathbf{H},Y)\ \mid\ a,a'\in A\ \}$.

References

Baer, R.: Free sums of groups and their generalizations III. Amer. J. Math. **72**, 647-70 (1950)

Karrass, Pietrowski, Solitar: Finite and infinite cyclic extensions of free groups. J. Australian Math Soc. (1973)

Rimlinger, F.S.: A subgroup theorem for pregroups. Proceedings of the Combinatorial Group Theory and Topology. Ann. of Math. Studies. To appear.

Serre, J.P.: Trees. Springer-Verlag. (1980).

Stallings, J.R.: A remark about the description of free products of groups. Proc. Cambridge Philos. Soc. **62**, 129-134 (1966a).

Stallings, J.R.: On torsion-free groups with infinitely many ends. Ann. Math. **88** 312-32 (1968)

Stallings, J.R.: Group theory and three-dimensional manifolds. Yale Monographs **4**, (1971).

van der Waerden, B.L.: Free products of groups. Amer. J. Math. **70**, 527-528 (1948).

General instructions to authors for PREPARING REPRODUCTION COPY FOR MEMOIRS

For more detailed instructions send for AMS booklet, "A Guide for Authors of Memoirs." Write to Editorial Offices, American Mathematical Society, P. O. Box 6248, Providence, R. I. 02940.

MEMOIRS are printed by photo-offset from camera copy fully prepared by the author. This means that, except for a reduction in size of 20 to 30%, the finished book will look exactly like the copy submitted. Thus the author will want to use a good quality typewriter with a new, medium-inked black ribbon, and submit clean copy on the appropriate model paper.

Model Paper, provided at no cost by the AMS, is paper marked with blue lines that confine the copy to the appropriate size. Author should specify, when ordering, whether typewriter to be used has PICA-size (10 characters to the inch) or ELITE-size type (12 characters to the inch).

Line Spacing – For best appearance, and economy, a typewriter equipped with a half-space ratchet – 12 notches to the inch – should be used. (This may be purchased and attached at small cost.) Three notches make the desired spacing, which is equivalent to 1-1/2 ordinary single spaces. Where copy has a great many subscripts and superscripts, however, double spacing should be used.

Special Characters may be filled in carefully freehand, using dense black ink, or INSTANT ("rub-on") LETTERING may be used. AMS has a sheet of several hundred most-used symbols and letters which may be purchased for $5.

Diagrams may be drawn in black ink either directly on the model sheet, or on a separate sheet and pasted with rubber cement into spaces left for them in the text. Ballpoint pen is *not* acceptable.

Page Headings (Running Heads) should be centered, in CAPITAL LETTERS (preferably), at the top of the page – just above the blue line and touching it.

LEFT-hand, EVEN-numbered pages should be headed with the AUTHOR'S NAME;

RIGHT-hand, ODD-numbered pages should be headed with the TITLE of the paper (in shortened form if necessary).

Exceptions: PAGE 1 and any other page that carries a display title require NO RUNNING HEADS.

Page Numbers should be at the top of the page, on the same line with the running heads.

LEFT-hand, EVEN numbers – flush with left margin;

RIGHT-hand, ODD numbers – flush with right margin.

Exceptions: PAGE 1 and any other page that carries a display title should have page number, centered below the text, on blue line provided.

FRONT MATTER PAGES should be numbered with Roman numerals (lower case), positioned below text in same manner as described above.

MEMOIRS FORMAT

It is suggested that the material be arranged in pages as indicated below.
Note: Starred items (*) are requirements of publication.

Front Matter (first pages in book, preceding main body of text).

Page i – *Title, *Author's name.

Page iii – Table of contents.

Page iv – *Abstract (at least 1 sentence and at most 300 words).

**1980 Mathematics Subject Classification (1985 Revision).* This classification represents the primary and secondary subjects of the paper, and the scheme can be found in Annual Subject Indexes of MATHEMATICAL REVIEWS beginning in 1984.

Key words and phrases, if desired. (A list which covers the content of the paper adequately enough to be useful for an information retrieval system.)

Page v, etc. – Preface, introduction, or any other matter not belonging in body of text.

Page 1 – Chapter Title (dropped 1 inch from top line, and centered).

Beginning of Text.

Footnotes: *Received by the editor date.
Support information – grants, credits, etc.

Last Page (at bottom) – Author's affiliation.

ABCDEFGHIJ – 8987